S0-AHD-770

THE POLITICS OF ZOOS

OF A NATION
CAN BE Judged
By The WAY
Its ANIMALS
ARE TREATED"
-GANDHI
BOB
DON'T
KNOW
EXTRA QUALITY & GROSS

The Politics of Zoos

EXOTIC ANIMALS AND THEIR PROTECTORS

JESSE DONAHUE AND ERIK TRUMP

NORTHERN
ILLINOIS
UNIVERSITY
PRESS
DeKalb

Published by the Northern Illinois University Press, DeKalb, Illinois 60115

Manufactured in the United States using acid-free paper

Design by Julia Fauci

Library of Congress Cataloging-in-Publication Data

Donahue, Jesse.

The politics of zoos: exotic animals and their protectors / Jesse Donahue, Erik Trump.

p. cm.

Includes bibliographical references and index.

ISBN-13: 978-0-87580-364-7 (library ed. : alk. paper)

ISBN-10: 0-87580-364-4 (library ed. : alk. paper)

ISBN-13: 978-0-87580-613-6 (quality pbk. : alk. paper)

ISBN-10: 0-87580-613-9 (quality pbk. : alk. paper)

1. Zoos—Political aspects—United States. 2. Animal rights—United States.

3. Animal welfare—United States. I. Trump, Erik. II. Title.

QL76.5.U6D66 2006

590.73—dc22

2006001580

Frontispiece is a detail from the photograph shown on page 51. Courtesy of the *Milwaukee Journal,* January 6, 1991. Photo by Richard Wood © 2005 Journal Sentinel, Inc., reproduced with permission.

To Maia and Miles

Contents

Acknowledgments

Although the task of writing this book required long, solitary hours in front of a computer, the truth is that our project benefited greatly from the advice of numerous people in a variety of disciplines. To a large extent, the most enjoyable portions of the writing process included our communications with these people.

Close to home, we received wonderful support from friends and colleagues. Attorneys Laura and David Donoghue provided detailed feedback on our early efforts to write about legal cases and also helped us locate court documents. Seth Gilbertson offered his expertise on legal footnoting. At Saginaw Valley State University, undergraduates Danielle Lautner and Melissa Stacer assisted us throughout the research process. Several of our colleagues read portions of the manuscript, including Robert Lane, David Weaver, Mark Nicol, Doug McGee, Julie Lynch, and Francis Dane. We are thankful to Dean Donald Bachand and Academic Vice President Robert Yien for granting the sabbaticals that gave us the time to start this book.

We particularly wish to thank colleagues from outside our university—including Kay Schlozman, Sidney Verba, Dennis Hale, Mark Landy, and Daniel Wirls—for reading early versions of the book and asking us thought-provoking questions.

We also owe a debt of gratitude to the many members of the zoo community who granted us off-the-record interviews and encouraged us to write about zoo politics. First, the American Zoo and Aquarium Association allowed us to attend both their legislative and annual conferences in 2003, where we made important contacts, including with Elin Kelsey, whose comments on an early draft helped us focus our manuscript. Other AZA members, who wished to remain anonymous, provided us with useful background information about the zoo community. Robert Wagner graciously answered our several e-mail inquiries. Craig Dinsmore generously allowed us to outbid him at an AZA Conservation Endowment Fund auction for a nearly complete set of *AAZPA Newsletters* for the period under study. We appreciate the friendly welcome we received from these zoo professionals.

Librarians and archivists also helped us with this project. Tom Zantow and David Shirley were particularly helpful in tracking down government documents and other sources that we required. We would like to thank the Smithsonian Institution and the AZA for granting us access to the archives that inform so much of this history. At the Smithsonian Institution Archives, Keith Gorman and Ellen Alers proved extremely helpful.

The task of locating illustrations for the book was made easier by generous assistance from Steve Johnson and Suzanne Balduc, at the Wildlife Conservation Society; Mary Stella, at the Dolphin Research Center; Patricia DeFrain, at the Milwaukee Public Library; Yedida Soloff, at the *Chicago Tribune;* Dennis Tennant, at the *Virginia Daily Press;* Raegan Carmona, at the Universal Press Syndicate; and Michael Polgany, at the Columbus Zoo.

We are grateful to the anonymous readers who read and commented on our manuscript for Northern Illinois University Press. Finally, and most important, we are indebted to Melody Herr, our editor, for her early and unflagging enthusiasm, as well as for her thorough editorial guidance.

Abbreviations

AAZPA	American Association of Zoological Parks and Aquariums
AHA	American Humane Association
ALDF	Animal Legal Defense Fund
ALF	Animal Liberation Front
APHIS	Animal and Plant Health Inspection Service (U.S. Department of Agriculture)
API	Animal Protection Institute
AWA	Animal Welfare Act (1970)
AWI	Animal Welfare Institute
AZA	American Zoo and Aquarium Association
CAB	Civil Aeronautics Board
CEASE	Citizens to End Animal Suffering and Exploitation
CHL	Committee for Humane Legislation
CITES	Convention on International Trade in Endangered Species of Wild Fauna and Flora
Defenders	Defenders of Wildlife
ESA	Endangered Species Act (1973)
ESCA	Endangered Species Conservation Act (1969)
FFA	Fund for Animals

FWS	U.S. Fish and Wildlife Service (Interior Department)
HSUS	Humane Society of the United States
IDA	In Defense of Animals
ILAR	Institute of Laboratory Animal Resources
IUCN	International Union for Conservation of Nature and Natural Resources
MMPA	Marine Mammal Protection Act (1972)
NMFS	National Marine Fisheries Service
NOAA	National Oceanic and Atmospheric Administration
NRPA	National Recreation and Park Association
NYZS	New York Zoological Society
OSA	Office of Scientific Authority
OES	Office of Endangered Species (Interior Department)
PAWS	Performing Animal Welfare Society
PETA	People for the Ethical Treatment of Animals
SAPL	Society for Animal Protective Legislation
SEIF	Special Education and Information Fund
SSP	Species Survival Plan
UAA	United Action for Animals
USDA	United States Department of Agriculture
WAPT	Wild Animal Propagation Trust
WWF	World Wildlife Fund
ZooAct	Zoo Political Action Committee

THE POLITICS OF ZOOS

INTRODUCTION

The Political Revolution

In 1919 twelve-year-old Blanche Guzzi and two of her friends decided, unwisely, to play ball near a bear cage at the Bronx Zoo. Blanche was standing closest to the bear enclosure when their ball rolled inside. Paying no attention to the bear, Blanche crawled through a protective fence and, lying down in front of the cage, reached in to get the ball. The bear, which had been watching the whole scene, reached over and grabbed the girl's head, tearing—and permanently injuring—her scalp.

Blanche's father sued the New York Zoological Society (NYZS), a private local interest group made up of elite citizens who served as the Bronx Zoo's governing board. Mr. Guzzi sought to recover for personal injuries sustained in the attack, claiming that "the keeping of the bear" that injured Blanche "constituted a nuisance." One of the few ways that American cities regulated animal ownership, including zoos, was through nuisance ordinances that punished animal owners for a variety of unpleasant side effects on the community, such as allowing domestic livestock to roam throughout the streets or failing to stop dogs from barking all night. This case tested the zoo's right to exhibit animals, particularly dangerous animals, because Guzzi was not claiming that the zoo had failed, by virtue of a poorly designed enclosure, to provide adequate protection from the animals. He was simply arguing that the very presence of bears at the zoo was a nuisance. By this logic, many other dangerous animals, such as tigers and lions, that drew so many patrons were nuisances as well. The court, however, found that the zoo served the public interest because it was governing an institution "chartered by the Legislature" and did not "maintain the animals for private gain." Therefore, bears could not constitute a menace. The court explicitly contrasted this public purpose with the private ownership of animals. "The liability of one who exhibits animals for private gain, differs from that of an institution maintained as a public enterprise." Thus, Mr. Guzzi could not recover damages because the zoo was not maintained for "private gain" but was a "chartered zoological society" that existed "for educational purposes and to entertain the public." The court thereby signaled that zoos ought to engage in education rather than simply entertainment because the line between a private individual who owned animals that he exhibited to the public for economic reasons and city zoos that existed solely for entertainment reasons was fairly thin. Education, coupled with sound fencing construction that kept animals from escaping, meant for this court that the zoo "could not possibly be guilty of maintaining a nuisance under the situation."[1]

In 1992, seventy-three years later, the NYZS was in court again, but under very different circumstances. The case pitted the Animal Protection Institute of America (API) as well as a host of other environmental and animal protection groups, including the Humane Society of the United States, against the Secretary of Commerce, Robert Mosbacher, in a bid to invalidate permits that he had granted to the John G. Shedd Aquarium (Chicago) for the purpose of capturing and displaying several whales protected by the Marine Mammal Protection Act of 1972 (MMPA). Quite different in its details from *Guzzi,* this case revealed the enormous revolution in the political landscape of zoos that had taken place in the two decades since the MMPA became law.

The case consolidated several lawsuits filed by ten animal protection organizations to prevent the Shedd Aquarium from capturing four beluga whales in the Hudson Bay and importing six false killer whales from Japanese aquariums.[2] To obtain these animals for its planned $43 million expansion, the Shedd had secured permits from the commerce secretary, as required by the MMPA. The central plaintiff, the API, disliked the fact that the MMPA gave aquariums a special privilege to collect wild animals that fishermen did not have, because, in their view, aquariums were just another kind of "fishing" industry that depended upon regular wild takings to remain commercially viable.

The API could not attack the MMPA directly, so it sought to protect the whales by contending that the permits violated MMPA requirements because the secretary had neglected to discover important facts related to the health of the animal populations in general and the characteristics of the particular beluga whales to be taken. Specifically, the API argued that Mosbacher had failed to determine whether the specific animals the Shedd planned to capture were "pregnant at the time of taking," "nursing at time of taking," or "less than eight months old." Instead, the permit simply prohibited the Shedd from taking an animal from one of the restricted categories, because it would have been nearly impossible for the aquarium to determine during the application process the *exact* marine mammals that it planned to collect. In practice, a pod of beluga whales would be herded into shallow water where they could be physically examined (by inserting a hand into the genital slit) and the few permitted whales would be culled.[3] The API and other animal protectionists hoped that, if the court agreed that herding an entire pod of whales was itself a "taking" (or at least "harassment," also prohibited by the MMPA), the Shedd and other aquariums would effectively be prevented from taking marine mammals at all.

Aquarium directors took this threat seriously, as indicated in a 1988 American Association of Zoological Parks and Aquariums (AAZPA) board of directors meeting at which they expressed "considerable concern" about the issue and again in 1991, when they worried that other animal welfare organizations sought "to end altogether the display of marine mammals."[4] (The AAZPA changed its name to the American Zoo and Aquarium Association in 1994 and shortened its abbreviation to AZA.) Recognizing that their interest in collecting the animals was directly at stake, the Shedd Aquarium and the AAZPA supported Mosbacher by petitioning to become intervener defenders. More than 50 percent of the AAZPA's member institutions held marine mammals in their collections and, together, made nearly twenty applications per year to obtain such animals. Any challenges to the permitting process potentially affected a significant portion of the AAZPA member institutions.

One of those members, the NYZS submitted an amicus brief on behalf of the aquarium, describing its own use of beluga whales in the public interest to advance conservation research and public education. Stressing its scientific knowledge about the display and care of marine mammals, the AAZPA reminded the court that it was "recognized by the government in the regulatory process as an expert on species protected under the law." It promised to "provide the court information about the public display and captive propagation programs that its member zoos and aquariums have undertaken" so that the court would understand why changing the permitting process as requested by the API in this case would hurt its member institutions. The AAZPA wanted to both help a member institution with its claims to superior authority on the topic and safeguard its other members in case the Shedd Aquarium inadequately defended the current permitting process.

Although the API clearly wanted to restrict the Shedd Aquarium's display plans, it was almost certainly destined to lose the case because of legislative protections for aquariums enacted in the 1970s. Specifically, the MMPA allowed industries to take "an inconsequential quantity of animals" for beneficial purposes, including public display. The MMPA established the Marine Mammal Commission (MMC) and the Committee of Scientific Advisors on Marine Mammals (CSAMM) to evaluate requests to take marine mammals, and in this case neither the MMC nor the CSAMM believed that the Shedd request would harm the wild populations of either species. Recognizing the API's strategy of trying to make animal welfare policy through the courts, the

Shedd's attorneys suggested that, if the API's attorneys wished "a change in the agency's regulatory regime," they were "free to request if from Congress. But they should not seek to use this Court as a cat's paw to pull their chestnuts from the fire." The court agreed with the interveners and found for the Secretary of Commerce and the aquarium. It noted that the MMPA had implemented a permitting process to prevent the "indiscriminate or at-will seizures of marine mammals," but it allowed aquariums that complied to collect and display such animals.

Quite different in its details from *Guzzi,* this case revealed the enormous revolution in the political landscape of zoos that had taken place. In 1919, the Bronx Zoo existed in a relatively simple political environment unregulated by the federal government, minimally concerned about conservation or animal welfare questions, and capable of resolving problems locally and without input from national interest groups. Much had changed since 1919, and most of the significant modifications in the political landscape of zoos had taken place since the late 1960s. The central questions that motivate this study, however, are how and why did these changes occur? And how did they affect the mission and management of America's professionally managed zoos and aquariums?

Visitors to zoos today recognize changes in zoo exhibits. Small, concrete cages have given way to large, naturalistic habitat enclosures. Animals are rarely displayed alone and are often accompanied by their offspring. Endangered species are noted in informational signage, and zoos generally participate in breeding programs to increase their numbers. Millions of Americans attend zoos today, but they are generally unaware of the politics behind these changes. They know quite a bit about the changing physical face of the zoo, but much less about the interest groups and governmental forces driving this transformation. Whether they realize it or not, what they are witnessing is the result of clashes between zoos and the interest groups that monitor them.

We examine the modern history of the zoo community as a political player in battles to protect captive and wild animals. Ours is a story about how dynamic individuals created an organization—the AAZPA—to represent zoos. We examine how these people reacted to newly energized animal welfare and animal rights activists from the 1960s forward.[5] We trace the political development of the AAZPA as zoo directors Robert Wagner, Theodore Reed, Ronald Reuther, William Braker, William Conway, and Gary Clarke, among others, remade it from a relatively small and powerless interest group into an organization that quite effectively represented zoos' interests. We examine what these leaders said and did *privately* to ensure that zoos could continue to collect and exhibit wild animals. We investigate largely unexamined government documents, archived materials, and AAZPA newsletters to uncover the staggering amount of political work that zoo and aquarium professionals engaged in from the 1960s forward.

What emerges from these documents is a Byzantine world in which zoo animals are at the center of political battles. Zoos have always been a part of the political system in which they reside: kings and conquerors around the world have exchanged animals with one another as symbols of power and alliances for centuries. The political history that we unveil, however, is a dramatically more complicated one that includes representatives of animal welfare groups, zoo directors, celebrities, members of Congress, regulators, judges, local politicians, and everyday citizens, all whom feel they have the democratically endowed right to determine zoo practices. We unearth a changing cast of animal protection advocates, some of whom wanted to abolish zoos completely and others who wanted to reform them. At the same time, Wagner and other zoo directors who ran the AAZPA had to find a way to protect their member institutions long enough to reform them. He had to discipline zoo directors while also convincing them to join the organization. And he had to translate his members' often idealistic animal welfare and conservation agendas into a practical program that included all zoos. Our story begins a bit earlier, however, in the parallel emergence during the 1950s of national animal welfare organizations and of a highly educated, conservation-minded group of zoo professionals.[6]

American animal welfare organizations had existed since the latter 1800s, when Henry Bergh created the Society for the Prevention of Cruelty to Animals, but none of the early organizations made zoo animals a central concern. In the 1950s, however, two interest groups emerged and eventually challenged zoos directly. In 1951, Christine Stevens formed the Animal Welfare Institute (AWI) to work against cruelty toward animals in research laboratories. Four years later, she formed the Society for Animal Protective Legislation (SAPL), which lobbied for laws that directly affected zoos. A tireless advocate for wild animals, Stevens frequently sparred with AAZPA leaders at congressional hearings and in the media. The Humane Society of the United States (HSUS), the AAZPA's most persistent critic, was created in 1954 by dissatisfied members of the American Humane Association (AHA) who wanted to address a variety of animal welfare issues involving agriculture, wild animals, and laboratory research animals. In one of the first acts that heralded its long-range political work, the HSUS moved its headquarters to Washington, D.C., to emphasize its self-identity as a *national* organization and more effectively monitor and influence federal actions regarding animals. Sue Pressman, a former Franklin Park Zoo (Massachusetts) employee, spearheaded much of the HSUS's work involving captive wild animals and interacted on a first-name basis with the AAZPA leadership during the 1970s and 1980s.[7]

In the meantime, the zoological profession was also changing. Originally conceived as an essential component of a comprehensive park system, zoos had long been funded by local taxes and managed by local parks department personnel. This location in the political system meant, effectively, that many zoo jobs were patronage jobs.[8] As a result, by all accounts zoo animals suffered from less-than-professional care. What changed in the 1940s and 1950s was that a large number of college-educated biologists and zoologists began to take jobs at zoos. The professionally and conservation-minded members of this group assumed leadership roles in major zoos and aquariums around the country, as well as in the AAZPA, the professional association that had represented zoos and aquariums since 1924 as a branch of the American Institute of Park Executives until 1966, when it became affiliated with the National Recreation and Park Association (NRPA). Increasingly dissatisfied with the NRPA's legislative and professional services, this small but influential group of zoo directors urged their fellow AAZPA members to create an independent organization in 1972. Putting up $25 apiece, nine directors established the AAZPA's initial treasury. Interest groups are typically run by "a fraction of a minority," and the AAZPA was no exception.[9]

A small cohort of directors from prominent zoos and aquariums guided the AAZPA through the politically turbulent period of the 1960s and 1970s, establishing a conservation and animal welfare identity for zoos that was reinforced in subsequent decades. Each of these key directors served a term as the AAZPA's president as well as a member of the board of directors and chaired committees that grappled with the most pressing political issues of the period. The senior member of this group, Theodore Reed, graduated with a degree in veterinary medicine from Kansas State College in 1945, served as assistant state veterinarian in Oregon during 1946–1948, and acted as the longtime director of the National Zoo (1956–84). Reed chaired committees on importation, legislation, and membership, and he played a key role in developing a national captive propagation program for zoos. William Conway, director of the Bronx Zoo (1962–99) earned an A.B. in zoology from Washington University, St. Louis. Pushing for a conservation role for zoos, Conway chaired a number of committees devoted to this issue. Ronald T. Reuther took his undergraduate degree in wildlife management from the University of California—Berkeley in 1951. Director of the San Francisco Zoo, Reuther served as AAZPA president from 1968 through 1970 and was a forceful advocate for the AAZPA's break from the NRPA. William Braker, the influential director of Chicago's Shedd Aquarium, received a bachelor of science degree from Northwestern in 1950 and a master of science from George Washington University in 1953. As AAZPA president (1973–74), Braker pioneered the association's use of lawyers in the battle with recalcitrant government agencies;

he also laid the groundwork for accepting private zoos into full membership. Gary Clarke, director of the Topeka Zoo (Kansas) from 1963 to 1989, held the AAZPA presidency during its first year of independence, overseeing the association's establishment of an accreditation commission and a range of new policies and standards for its members.[10]

By far the most politically influential figure in the AAZPA, the boundlessly energetic Robert Wagner studied zoology at Purdue University in the early 1950s and then worked his way up the zoo ladder to become director of the Washington Park Zoo (1959–64), in Portland, Oregon, and the Jackson (Mississippi) Zoological Park (1964–75). One of the AAZPA incorporators, Wagner served on the board and chaired its legislative committee before being chosen executive director in 1975, where he remained until 1992. During his tenure, the AAZPA grew dramatically from a tiny organization with few members to one with a multimillion dollar budget.

Despite their commitment to protecting wild animals, all of these AAZPA figures became, at times, formidable opponents of animal welfare and rights activists in the 1970s, 1980s, and 1990s. One of the central ironies of our story, moreover, is that, just as animal welfare organizations began to argue that zoos and aquariums did not know how to take good care of animals and did not care for the well-being of wild species, the opposite was true. Zoos and aquariums were attracting well-educated professionals who cared a great deal about what happened to species in the wild and at their institutions.

The political battles between the AAZPA and an increasing number of animal protection groups animate much of the story that follows. Our history analyzes the behavior of the AAZPA and its member zoos from the 1960s to the present, stressing particularly important events within chapters to highlight their influence on the political development of zoos and aquariums. Within this chronological narrative, several themes recur.

First, the AAZPA's leaders had to clarify and define the association's mission, including who would be accepted as members. These leaders adopted a conservation and educational mission. Because they had an economic and philosophical interest in preserving endangered species in the wild, they consistently supported a range of environmental legislation, even when that legislation promised to burden zoos with additional regulations. At the same time, they addressed internal conservation issues by developing captive breeding programs for endangered species. The AAZPA's official conservation mission was repeatedly complicated by the fact that their member zoos and aquariums were institutions that combined functions. They were educational, commercial, public, and recreational all at the same time. The lack of singular purpose, however, led to problems for zoos that they struggled to resolve during the entire period that we cover. Although the *Guzzi* case signaled that zoos were safer if they were public institutions, the AAZPA included a substantial minority of privately owned zoos and aquariums, often quite wealthy. A question that faced the organization early on was whether it should give private zoos and aquariums full membership rights. As an interest group, the AAZPA had a substantial incentive to include private, profitable zoos and aquariums that could contribute to the financial health of the group and thereby fight back against animal protection groups that were, ironically, accusing them of using animals for commercial gain. On the other hand, the financial motivation of private zoos somewhat compromised claims of performing a public service. The AAZPA decided that the benefits provided by commercial members were worth the risk, but the presence of these commercial institutions provoked conflicts with legislators, government agencies, animal protectionists, and even among AAZPA members.

Second, Wagner and the other key AAZPA members learned how to defend zoos' interests in various political arenas. They had to organize and educate their members about the political system, hire professional lobbyists, and adopt flexible political strategies to protect themselves and their animals against a variety of legislative, judicial, and interest group threats. As the AAZPA's executive director, Wagner in particular had to find a way to represent coherently and reasonably zoos when, at times, his fellow directors offered contradictory, idealistic, and even naive political advice. Wagner, together with hired consultants and a handful of active directors, had to negotiate the American political architecture that provides citizens multiple points of access through state and federal legislatures, agencies, and courts. When animal protection groups lost battles in the legislature they simply took advantage of the porous nature of the governmental structure and tried to have their policy goals implemented through the administrative or judicial branches.

Third, the AAZPA and individual zoo directors repeatedly reached out to form alliances with other interest groups and public agencies.[11] Zoo politics involved more than just legislative and legal squabbling between adversarial interest groups. Almost from the beginning of its existence, the AAZPA engaged in environmental protection work and tried to ally with potentially friendly conservation and animal protection groups. In some circumstances the AAZPA also quietly sought shelter behind more powerful agricultural and research interest groups that could more aggressively attack animal protectionists. At the administrative level, Wagner and other directors worked on winning over both the Department of the Interior and the Department of Agriculture. The AAZPA's relationship with both agencies was not always smooth, because animal protection groups also struggled to shape each agency's regulation of zoos. Because new federal regulations required zoos to upgrade their facilities, the AAZPA also encouraged member zoos to develop public-private partnerships with local zoo support organizations as a means of securing greater funding and thereby increasing the quality and number of both their exhibits and employees.

Finally, the AAZPA leadership had to strengthen the association's internal organization in order to unite and police its own members. Although the AAZPA's membership voted overwhelmingly for independence in 1972, the association represented a wide variety of members with differing interests and expectations. To become an effective interest group, the AAZPA required an organizational structure with clear leadership roles, as well as a set of mandatory guidelines for both institutional and individual members. From the beginning, however, the professional standards, accreditation guidelines, and political strategies preferred by the AAZPA leadership were not always embraced by the full membership. To address this lack of unity, the AAZPA first encouraged and then required compliance with uniform standards governing the operation of zoos, particularly the handling of endangered species. Still, internal debates about lobbying strategies, conservation policy, surplus animals, and animal protectionists continued throughout the period under study here.

Our history begins in the early 1960s, a period of optimism and new frontiers for Americans. Guided by a cohort of ambitious young zoo professionals, the AAZPA began to take positive steps to address a growing international conservation crisis. The world's supply of wild animals was dwindling as a result of habitat destruction and commercial exploitation, and the AAZPA's leadership believed that zoos could do something to stem the loss. By the end of the decade these same leaders stood in shock as it became clear that some Americans perceived zoos as part of the problem, not the solution.

ONE

Opening Moves

Protecting Animals from Indifference

In 1963, Fairfield Osborn, president of the New York Zoological Society, wrote an article for the *New York Times* titled "Another Noah's Ark—For New York." Fairfield reviewed the Bronx Zoo's efforts to develop captive breeding programs for endangered animals and proposed that zoos would become the Noah's Ark of the modern world as habitat destruction eliminated species around the world. This metaphor—of the zoo as a savior of endangered animals—came to justify zoos' existence in the decades ahead. Because habitat destruction and uncontrolled hunting threatened many species with extinction, it seemed reasonable to propose not only that zoos could harbor the few remaining representatives of a species, but also that successful captive breeding programs might one day enable the reintroduction of extinct species into the wild. In light of the potential of captive breeding to help endangered species, zoo professionals could logically argue against public policy that would entirely prevent their institutions from collecting, exchanging, or exhibiting such species. Thus, in the 1960s no one within the American Association of Zoological Parks and Aquariums perceived zoos as a significant threat to endangered species. In 1966, for example, the AAZPA's Conservation of Wildlife Committee reported that it "knew of no species on the endangered list which reached that unhappy state because of us. By far the chief threat to wild species is conversion of habitats to other uses."[1] The AAZPA was likely correct in this assessment, but the decade of the 1960s would see zoos' positive actions on behalf of endangered species swamped by a broader cultural and political concern about the plight of animals, and by the decade's end critics often identified zoos as part of the problem, not the solution.

The AAZPA's leaders saw their organization laboring internationally and domestically to get governments to provide better protections of wild animals, particularly of endangered species, and in fact zoo critics tended to underemphasize zoos' conservation and animal welfare activities during this period. The AAZPA participated in international conservation debates, developed captive breeding programs, imposed restrictions on its members' animal imports, and supported domestic endangered species legislation. Moreover, the AAZPA frequently criticized the federal government's failure to protect the welfare of zoo animals destined for the United States.

Zoo directors in the 1960s shared Gifford Pinchot's famed "wise-use" philosophy and applied it to animals. They were conservationists in the sense that they felt animals should be managed for the benefit of humans. To place zoo directors' conservationist philosophy and actions in perspective, it is important to understand that they were one of only a handful of wild animal conservation organizations in existence that were not hunting groups. They were joined only, for example, by Defenders of Wildlife (or Defenders) and the Audubon Society. It was only later that mainstream animal welfare groups such as the Humane Society of the United States (HSUS), the Animal Welfare Institute (AWI), the Animal Protection Institute, and Greenpeace, to name a few, actively began working for wildlife conservation.[2]

Rachel Carson, a founder of the U.S. environmental movement, held an animal conservation philosophy similar to that held by zoo directors. Carson was not in favor of the rights of wild animals, instead arguing at times that animals should be protected for the benefit of humans, including hunters. Similarly, she believed that it was all right to "trade off individual animals for the good of their species." Thus, zoo directors who formed the core of the AAZPA were, in the context of their time, animal conservationists.[3]

Even as zoo professionals pursued a conservation agenda, they found themselves shadowed by an animal welfare movement that was expanding its focus to include both wild animals and wild animals at zoos and building public and congressional support for increased federal regulation of zoos. Although these animal welfare organizations initially targeted small "roadside" zoos as the main offenders, all zoos were eventually affected by Congress's 1970 amendments to the 1966 Laboratory Animal Welfare Act (PL 89-544; AWA). AAZPA leaders ended the decade reeling from the realization that their self-image as conservationists was not widely shared by those outside the zoo profession and hurriedly taking the steps necessary to strengthen both their organization and its political position.

Wildlife Conservation Efforts

Throughout the 1960s, Theodore Reed, Gunter Voss, Ronald Reuther, William Conway, and other zoo directors from primarily nonprofit or public institutions used their leadership positions within the AAZPA to focus intently on wildlife conservation outside and inside zoos. Well educated and internationally connected, these zoo leaders recognized that the future of high quality zoos depended upon their enlightened engagement with conservation issues. The AAZPA could lead American zoos by both developing relationships with international conservation organizations and establishing guidelines to restrict its members' breeding and importation of endangered species. Although there was a degree of self-interest in these efforts—after all, zoos were in the business of exhibiting the animals that the AAZPA was trying to protect—a genuine concern for wildlife clearly motivated Reed, Voss, Conway, and their like-minded colleagues.

Because zoos imported so many of their animals from abroad, it made sense for the AAZPA to join international efforts to protect endangered species and habitats. In 1963, the AAZPA became an active member of the International Union for Conservation of Nature and Natural Resources (IUCN), an international body of scientists and environmentalists hoping to reverse some of the environmental damages humans had created. Several years later, Voss, as head of the AAZPA's conservation committee, proposed "broadening and deepening [their] working relationship with the principal conservation groups" so that they could cooperatively "build a global strategy" for threatened species. To a certain extent, AAZPA participation in such international conservation organizations helped zoo directors keep tabs on dwindling supplies of zoo animals, but it also reflected a genuine willingness to focus political attention on the plight of endangered species. Sometimes American zoo directors worked on species preservation efforts abroad that had no immediate benefit for their own institutions or other American zoos. At the Ninth General Assembly of the IUCN, in Lucerne, Switzerland (1966), for example, John Perry, an AAZPA member and assistant director of the National Zoo, was asked to help organize an international protest to stop the capture of Sumatran rhinoceroses in Malaya. It was becoming clear, however, that, even if wild populations remained stable, some foreign governments were "tightening [their] controls" on animal exports, and the AAZPA feared that many of those governments would see little reason to continue resupplying zoos. Other governments, however, took no actions to restrict the taking of wildlife. Reed, the chair of the AAZPA's conservation committee, summarized the situation in 1967 as "threatening . . . yet challenging. We are challenged to do more of the very things most of us want to do, and believe we should do. The question is whether we can do them soon enough."[4]

The inability of the international community to effectively stem the loss of endangered species persuaded the AAZPA to shift more attention to controlling the import behavior of its member zoos. The AAZPA began by blacklisting species such as the orangutan, monkey-eating eagle, Javan and Sumatran rhinoceroses, and Galapagos tortoise, prohibiting its members from importing these animals in the absence of any government laws to that effect. In 1962, eleven years before the Endangered Species Act (ESA) became law, it resolved that "all endangered species that are fully protected and adequately managed in their country, receive the additional protection which we can afford by our refusal to purchase any wild-caught specimens of these species which are removed illegally from country of origin and subsequently offered for sale on the animal market."[5] Although this resolution spoke only to "illegal" animal captures, it marked a first step in distinguishing zoos from the fur and pet industries, which not only imported far larger numbers of animals, but also were likely to employ poachers and other unsavory actors in their transactions. If zoos wished to be seen as biblical arks, they needed to stake out a moral high ground in their manner and purpose of collecting wild animals.

The AAZPA enjoyed no godlike power, however, and policing its own members' import activities turned out to be difficult. Although the board had blacklisted several species by 1966, some member zoos ignored the blacklist. In a few cases, the threat of a blacklist even had the opposite effect, prompting zoos to buy up animals before they would become unavailable. To address this problem, the board decided to establish a permanent subcommittee to review and rule on zoos' plans to import "legal" orangutans and other blacklisted animals. Such a review, though admirable in purpose, lacked much practical force, especially since the board chose not to involve the government and rejected one member's suggestion that offending zoos be publicly reprimanded. In addition, the conservation committee itself admitted that it lacked the "resources necessary" to monitor wild animal populations effectively—changing international laws—and multiple transactions between animal dealers; yet, without that information, the committee could not effectively evaluate zoos' acquisition plans for more than a handful of species. Fred Zeehandelaar, the AAZPA president's import advisor, had little patience for the board's toothless response to offending member zoos, pointing out the political dangers of inaction: "We are our own enforcement agents. We are to expel without warning any member who violates past resolutions or official regulations. . . . We cannot stand on our rights to import animals of any kind, through any Government agency, not only USDA but also others—on a legitimate basis if we do not clear our own ranks forthwith of violators." Three years later, the board congratulated itself for holding a "precedent-setting Endangered Species Resolution Violation Hearing" and handing out a one-year suspension to one of the two offending members, but the window for self-policing was closing.[6]

A commitment to worldwide conservation was critical for zoos' survival, because only a "small minority" of the mammal species in their collections reproduced in self-sustaining numbers in captivity, meaning that, unless their wild populations were protected, those species would eventually disappear from zoos. Zoo directors knew this because, beginning in 1963, the U.S. Fish and Wildlife Service (FWS) had conducted an annual census of members' animals to predict how those captive populations would be affected by breeding and the decline of wild populations. Its 1966 census report noted that the FWS "recommends for several species [as well as for some nonrepresented species], that zoos acquire additional breeding stock and intensify captive propagation."[7] Both the FWS and the AAZPA recognized that gains in scientific knowledge made it likely that a much larger proportion of the captive species could become self-sustaining in the future, but in the short term zoos would need to take additional specimens from the wild for breeding stock.

Not surprisingly then, part of the AAZPA's strategy for protecting animals included breeding them for zoos. The AAZPA spent much time in the latter 1960s discussing propagation plans that anticipated the Species Survival Plans created more than a decade later. It participated in an international animal conservation group called the Wild Animal Propagation Trust (WAPT) that included a variety of key zoological institutions such as the San Diego and Bronx zoos. This group's primary function was to coordinate the captive propagation of endangered species in zoos. Members of the AAZPA's various conservation committees and subcommittees also served simultaneously in WAPT and the IUCN, ensuring a great deal of coordination between the groups. For example, Reed served as president of WAPT and chairman of the AAZPA's conservation committee; two committee members also served on the IUCN's Survival Service Commission. In 1966 Voss, fearing that IUCN and Brazilian actions to protect the golden marmoset "may be too late," recommended that the board work with WAPT to set up breeding programs for the animals in American zoos; the board approved this motion and further determined to urge the Brazilian government to prohibit the export of golden marmosets. In response to this critical endangerment of golden marmosets, the board soon recommended that its members place the animals only in zoos where they could be bred. Similarly, in an effort to preserve orangutans in the wild and for display, the AAZPA cooperated with the FWS enforcement division to police importations and encouraged its member zoos to participate in exchanges to maximize captive breeding opportunities. As the conservation committee implored, "[I]t is urgently necessary that breeding take precedence over exhibition aims."[8]

By 1967, the AAZPA had adopted a formal resolution advocating the "judicious collection and placement of specimens in zoos or other centers capable of propagating these species in captivity." The resolution further promised that AAZPA members would "impose responsible self-discipline upon themselves and . . . cooperate with international conservation groups and zoo associations in other countries" to prevent the importation of several critically endangered animals; finally, the resolution contained a mechanism whereby additional species could be added to the blacklist.[9] Language also committed the AAZPA to developing close working relationships with the IUCN, foreign governments, and conservation organizations around the world.[10]

This voluntary approach to the responsible importation and management of endangered animals helped the AAZPA distinguish its members from commercial, less-conservation-minded importers. Sometimes, other governments that were "disillusioned" with such commercial importers and dealers called upon the AAZPA to provide them with "advice and help in regulating trade in animals." In 1967, Reed reported that the Thai government had asked the AAZPA to check the validity of permits for research primates. That same year, the Kenyan government requested information about which zoos and research institutions in the United States were legitimate, and an Indonesian government representative met with the AAZPA and WAPT to discuss how his government should handle the export of orangutans.[11] The AAZPA welcomed these consultations because they spoke to the organization's credibility and importance. It worried, at times, however, that other groups' advice could be considered more important. Its response to the ESA and its precursors illustrates how a philosophical support for endangered species legislation could conflict with a desire to be seen by the U.S. government as a foremost authority on wild animals.

Domestic Endangered Species Legislation

The AAZPA reacted favorably to two conservation bills that predated the ESA (1973): the Endangered Species Preservation Act (1966) and the Endangered Species Conservation Act (1969). Together, these two acts provided for limited habitat protection, authorized the creation of wildlife refuges, expanded the Lacey Act's prohibitions against the importation of and interstate trade in illegally taken animals and laid the groundwork for international restrictions on trade in endangered species by directing the Secretary of the Interior to develop a list of endangered species.[12] The Lacey Act, created in 1900, was one of the first pieces of environmental legislation that directly involved the federal government. It was designed to cut back on the killing of birds. The AAZPA had already recommended state-level protective legislation to save such animals as the Texas tortoise from extinction, so it naturally lent its support to these proposed federal endangered species acts. As the ESCA made its way through Congress, the AAZPA board discussed the proposed bill "at length" and agreed to "support the principle" to conserve endangered species. Always aware of zoos' "keen interest in obtaining animals with a minimum of difficulty," the AAZPA's leadership perceived no direct threat in this legislation and endorsed it unanimously.[13]

The proposed ESCA bill was reviewed by a central figure in AAZPA politics: Fred Zeehandelaar, an animal trader who served as an "import advisor" to the AAZPA president. Although some of the zoo directors who played large roles in the AAZPA during this period were certainly cosmopolitan and keenly aware of the role of politics, Zeehandelaar was one of the most politically savvy and cosmopolitan among them. An immigrant from the Netherlands, Zeehandelaar had operated a wild animal import service since 1957, serving zoos, research laboratories, and individuals. In this role, he became intimately familiar with international law and the U.S. agencies regulating animal importations, and his general view was that governments did too little to protect the wild animals he loved. Viewed suspiciously by the animal protectionists (who delighted in his indictment in 1972 for a violation of the ESCA), and despite his economic interest in being able to import animals, Zeehandelaar described the "wonderful conservation people" as a "special class of humans" who were trying to assure that wild animals would "have a very bright future."[14] A prolific letter and memo writer, he obsessively reviewed all legislation and regulations that might possibly impact both his own exotic animal importation business and zoos. Acknowledging his expertise on international regulations and legislation affecting zoos animals, the AAZPA sought his advice throughout the 1960s and into much of the 1970s. Reviewing an early version of the ESCA, Zeehandelaar concluded that as the "President's Import Advisor as well as personally in [his] own business" he was "100% in favor of the purposes of the proposed bill."[15] Other zoo leaders concurred, and representatives from both the AAZPA and WAPT testified in favor of the legislation at a 1967 House subcommittee hearing, accepting the bill's purpose of regulating traffic in endangered species but managing to "work out a common position with Department of Interior and other federal agencies" regarding some of their concerns about details in the proposed legislation.[16]

In part, the AAZPA's initial concerns about the ESCA reflected its desire to protect its interests. First, the board expressed concern that the bill did not "provide for an advisory committee" to oversee the Secretary of the Interior's decisions about which animals to place on the endangered species list. As the "greatest and largest nonprofit" organization in the field, the AAZPA felt that it should have a place on such an advisory committee. Although the bill did say that the secretary should consult "interested persons and organizations," the organizations it specified as examples were "ornithologists, ichthyologists, ecologists, herpetologists and mammalogists." This situation, from Zeehandelaar's perspective, caused "a danger, a great danger, of primary advice being sought outside of the U.S.A, specifically with foreign governments." His worries were confirmed when he asked the director of the Bureau of Sport Fisheries and Wildlife (later the FWS), "What is your order of value in asking advice on which animals should be declared endangered?" The Director replied, "(1) Foreign Governments (2) IUCN (3) AAZPA." This state of affairs Zeehandelaar pronounced "wrong." He argued that the AAZPA should come first because, unlike foreign governments, it was not influenced by "borderlines or political upheavals or hard currency requirements," each of which could influence a government to misrepresent the actual status of its endangered animals. In addition, the AAZPA's close ties with the IUCN meant that it could represent that organization's advice faithfully. The AAZPA desired some control over which species received endangered status because its members had such a direct interest in being able to obtain wild animals. On balance, however, even the creation of an endangered species listing did not worry zoo professionals such as Perry, who noted that such a list would likely include mainly species already blacklisted by the AAZPA and protected by law in their own countries. Pointing to the secretary's power to grant importation exceptions for "zoological purposes," as well as the fact that similar legislation in England had resulted in the listing of only forty species, Perry confidently concluded that the ESCA would not hurt zoos.[17]

In a broader sense, the AAZPA also wanted to make sure that the government recognized the value and legitimacy of zoos' emerging breeding programs and its proactive efforts to develop responsible importation guidelines for its members. Neither the 1966 nor the 1969 act added major new restrictions on zoos' ability to collect wild animals. The 1969 act did, however, empower the Secretary of the Interior to create a list of *globally* endangered species, prohibit the importation or interstate commerce of species in danger of becoming extinct, and work toward the creation of an international treaty restricting trade in endangered species (this became the Convention on International Trade in Endangered Species of Wild Fauna and Flora, the CITES treaty, written in 1973 and ratified by Congress in 1975). Significantly, neither the eventual ESA nor CITES provided for a "broad program of species conservation." Rather, they banned or restricted trade in those species.[18] Without habitat protection and other kinds of support for endangered species, there was no guarantee that restriction or even prohibitions on trade would actually result in population recoveries.

In the context of the proposed endangered species legislation's modest goals, the AAZPA could reasonably imagine that it was doing *more* than the government to conserve endangered species, since it was both restricting the numbers of animal species taken by zoos and increasing the number of animals raised in captivity. In fact, the AAZPA implored Congress to tighten up language in the ESCA to prevent just *any* zoo from being able to import endangered species. The draft legislation allowed the Secretary of the Interior to permit the importation of endangered species "for zoological" or "educational" purposes. Testifying for the AAZPA, Reed pointed out that not all zoos were suitable recipients of such animals and that only captive breeding, not education, constituted a legitimate reason for importing endangered species.[19] In short, the AAZPA lined up with conservation organizations, the Department of the Interior, and a nearly unanimous Congress to support the principle of restricting trade in endangered animals. On December 5, 1969, Richard Nixon signed Public Law 91-135, which authorized the Interior Department to identify endangered species and prevent the importation of such animals for all but "zoological, educational, scientific, or propagation purposes." Successful in their efforts to protect zoos in the legislative process, the AAZPA's leaders soon discovered the wisdom of Zeehandelaar's frequent admonition to pay close attention to the implementation of laws. Unfortunately, few within the AAZPA appeared to have developed close working relationships with Interior officials, perhaps because zoos had previously had little need for sustained contact with the agency. The absence of familiarity proved irksome at first, but turned dangerous by the early 1970s.

Initially, it appeared as though the ESCA would help the AAZPA police its members' importation practices. The AAZPA had had difficulty getting its members to honor blacklists, but the ESCA promised to solve that problem only if the Department of the Interior could be persuaded to share member zoos' endangered species import permit applications with the conservation committee. By reviewing those applications and making recommendations to Interior, the committee could effectively control AAZPA members' import practices. For nearly a year after the passage of the ESCA, however, Interior officials refused to allow the committee to review the applications, providing only copies of awarded permits. In 1971 Interior's Office of Endangered Species relented, giving the committee "its first opportunity to review and evaluate a zoo application." The committee hoped that this event would mark "the beginning of a long-awaited routine procedure by which the AAZPA will review each zoo application received by the USDI."[20] Interior did not cooperate as expected, and the committee managed to review only a few of the forty-four permits granted to zoos over the next year. As a way to "reinforce the cooperation between the USDI officials and the AAZPA," the Wildlife Conservation Committee chairman resurrected voluntary cooperation measures and recommended that zoos send a duplicate application to his committee, thereby allowing the committee to evaluate "the application before the USDI acts."[21] The committee did not simply rubber-stamp zoos' applications, as can be seen in its actions during 1973: of eight applications for the importation of wild-caught animals, the committee disapproved of three, resulting in permit denials from the USDI.[22]

The AAZPA's desire to review permit applications was a logical extension of its earlier attempts to blacklist certain endangered animals, and it points to the extent to which the AAZPA tried to insert itself into a quasi-official decision-making role and police its own members. On the one hand, the AAZPA wanted to protect its members' interests during the permitting process by providing recommendations to the USDI; on the other hand, the AAZPA did not want its members to violate the spirit or letter of its own endangered species policy. The AAZPA leadership was expending significant energy during the 1960s to promote and support a range of international conservation efforts, and it could not afford to have the positive conservation image of zoos tainted by the actions of uncooperative members. The conservation image of zoos would obviously be enhanced by developing a close working relationship with Interior, but Interior proved less cooperative than zoo leaders hoped, and its officials gave little indication that they planned to treat the AAZPA as a partner in decisions about importations or endangered species listings. No doubt this reluctance stemmed in part from the fact that the FWS staff included many trained scientists who reasonably believed that they possessed sufficient expertise to protect endangered species. Although there were moments of cordiality, cooperation, and mutual respect between AAZPA leaders and officials of the Department of the Interior, their conflicts over the importation and trade of animals intensified after the passage of the ESCA. Zoos' "wise-use" conservation philosophy fit uneasily with government officials who appeared increasingly skeptical about zoos' conservation credentials.

In 1968, Perry and others within the zoo world perceived little danger in the ESCA, focusing instead on how the legislation supported their own approach toward the responsible regulation of wildlife trade. Because zoo professionals, especially those in positions of leadership, perceived zoos as conservation organizations, they imagined that conservation legislation would help them achieve their goals of self-regulation. In the short term, zoos discovered that the ESCA instead precipitated a period characterized by often hostile encounters with Interior and bitter battles over unexpected rules restricting zoos' trade in endangered animals. In the long term, zoos realized that their role as conservation organizations would have to be argued vigorously and loudly, because it was not self-evident to all. Zoo leaders' early optimism regarding Interior may seem naive, but it likely reflected their contrastingly positive experiences with the United States Department of Agriculture (USDA), an agency with which zoos had long dealt and generally been able to work out its problems.

Working with the USDA to Save Animals from Government Neglect

While the AAZPA worked on international and national legislation aimed at saving wild animals, the American government treated zoo-bound animals with what sometimes seemed like callous disregard. Once zoos had secured the necessary import permits and had contracted with a shipper, their animals sometimes ended up dying in USDA quarantine facilities or, worse, euthanized by USDA officials fearful of exotic diseases. This situation deeply perplexed zoo professionals, who sometimes found the bureaucracy relatively unsympathetic to their concerns for animal welfare. Nonetheless, as several dramatic incidents suggest, USDA actions occasionally captured the public's attention, and the AAZPA employed all means it could to save the lives of animals caught in political controversies. In these actions, numerous sympathetic USDA officials assisted zoos, laying the groundwork for productive working relationships in the decades ahead.

The USDA oversaw exotic animal importation under authority of injurious animal legislation designed to protect the domestic livestock industry. Exotic animals imported from foreign countries had to spend time in designated quarantine facilities located around the country near ports of entry. The USDA sought mainly to ensure that imported animals, including zoo-bound exotic animals, did not carry diseases that might be transmitted to domestic livestock. Once in quarantine, exotic animals received care from USDA personnel most familiar with commercial breeds and not always aware of exotic animals' special needs. It is not surprising that zoo directors complained that too many of their animals were dying at one of the USDA's larger facilities, located at the Clifton Airport in New Jersey. A study by the USDA supported this accusation. Between 1959 and 1965, losses at the facility ranged from "a low of 3.33% to a high of 12.0% with an overall percentage for the seven years of 6.44%."[23] Such losses might be acceptable for commercial animals interests, but, because zoos generally imported only a few specimens of each species, the death of even a single animal could destroy breeding plans and necessitate a time-consuming search for a replacement.

By 1965, the AAZPA Importation, Exportation, Health, Welfare, and Quarantine Committee had begun to take steps to improve communications between zoos and USDA officials and jointly develop procedures for processing quarantined animals and dealing with serious disease outbreaks. First on the committee's agenda was stemming the too-frequent deaths at the Clifton quarantine facility. In a series of letters and phone calls, Frank M. Thompson, the committee's chairman, discovered that the Clifton facility's director, Dr. Waters, not only was receptive to zoo's concerns, but also expressed "pleasure" at the prospect of cooperating closely with the committee to prevent future deaths. Thompson pledged to invite Waters to the next AAZPA conference, and he viewed their relationship as the first in a series that would begin at the lower levels of the USDA hierarchy and work slowly toward the top.

Contact with higher-level officials would be necessary for the next item on the committee's agenda: getting the USDA to create an emergency plan for hoof and mouth disease outbreaks that would not endanger zoo populations. The committee recognized that the USDA had a legitimate need to control such outbreaks but also feared that the USDA might destroy all zoo livestock in the event of an outbreak, an action that would potentially result in the loss of the "last living representatives of endangered species." Thompson warned his committee members that developing the liaisons that could ensure a favorable policy would "require considerable attention," but he also reminded them that their "thorough job with our assigned projects" was of tremendous importance to the zoo profession.[24]

A well-publicized incident in 1966 dramatized the unintentionally cruel effects of USDA policy regarding animal imports and disease and illustrated the urgency of establishing close communication with USDA officials. The *Maaslloyd,* a Dutch freighter, stopped in the Kenyan port city of Mombasa to pick up a large shipment of zoo-bound animals, including some rare species and quite a few giraffes. USDA policy at the time required that shipped animals come straight to the United States to avoid contamination by stopping in other countries because they were concerned about a hoof and mouth disease outbreak at two ports in Africa. The freighter's captain, however, subsequently made an unscheduled stop in another port, whereupon the USDA instructed him to euthanize all animals on board and dump them at sea before entering the United States. News of the situation reached the media before the captain could comply, and pressure from the HSUS and other organizations convinced the USDA to quarantine the animals on a small island in Long Island Sound. Cartoonist Pat Oliphant recognized the politics of the incident and drew a picture of Noah and his wife on an ark observed by frightened animals while Noah asks, "I should take them outside the twelve-mile limit and dump them because they *may* have hoof and mouth disease?"[25]

To avoid this kind of disaster in the future, the AAZPA president (Don Davis), the president-elect (Conway), the National Zoo's Reed, and others arranged a meeting with USDA representatives, animal importers, and carriers, at which they raised several concerns such as providing greater care of animals during quarantine, using a qualified zoo veterinarian to evaluate imported exotic animals, separating zoo and domestic agriculture animals, and increasing the number of quarantine stations overseas. The immediate outcome of this meeting was that the AAZPA representatives convinced the USDA to test animals before they left their country of origin, fly the test to the United States, and, if the animals tested negative, allow them to come into the country. More important, however, the experience led to the creation in 1967 of a subcommittee designed to "act as consultants to the USDA" and help it "solve the problems" related to the importation and quarantine of zoo-bound animals. Conway cultivated a positive relationship with Dr. Frank J. Mulhern, the Deputy Administrator for Regulatory and Control, Agricultural Research Service, by assuring him of his "access" to the "Association's leadership" and expertise, congratulating him on promotions, commending him on reports of improvements at Clifton, and extending invitations for face-to-face meetings to "pursue our problems and opportunities for improvement." In short, Conway and the AAZPA leadership recognized the long-term value of being seen as friendly, helpful constituents of the USDA.[26]

Despite this cooperation, the USDA's deeply engrained focus on domestic, but not exotic, animals was apparent when it reorganized senior staff members in 1967. Each was given responsibility over a commercially significant section—horses, bovine, swine, poultry—but none was assigned to zoological animals. Zeehandelaar summed up the situation by arguing that "the powers to protect against introduction of foreign diseases are without limit, the powers to—under due protection—stimulate importation [of zoo animals] are hopelessly limited." From his perspective, it was the "first duty of the USDA to find a legal reason not to grant an import permit." Part of the reason, according to Zeehandelaar, was incomplete knowledge. Some of the officials, for example "place Antwerp in the Netherlands; others do not know the difference between a consul and a counsel . . . others confirm that the Kiang and the Tiang are one and the same animal."[27] In 1968, USDA officials at the Plum Island quarantine facility, in New York, decided that healthy, domestic cattle might be contaminated by healthy, vaccinated ruminants imported for zoos. This fear threatened to shut off zoos' supply of ruminants and was characterized by Zeehandelaar as a "ridiculous restriction, bordering on hysteria."[28] In light of the USDA's historical mission of serving domestic agricultural interests, it was hardly surprising that the agency had little expertise with and few resources to expend on zoo animals.

When zoo animals fell victim to the USDA's bureaucratic indifference, AAZPA members expended tremendous energy to save them and were assisted by their agency contacts. One incident involving Zeehandelaar illustrated the USDA's simultaneous indifference and cooperation when it came to zoo animals. On August 9, 1971, a shipment of roan, greater kudu, and springbuck antelopes arrived at New York from Walvis Bay, South West Africa. On Friday, August 27, Zeehandelaar received a call from the Animal Health Division of the Agricultural Research Service, informing him some of the animals' blood tests indicated that they might be carriers of a virus that could infect cattle. These test results were akin to a death sentence. Hoping to avoid a "forthcoming catastrophe and the black future," Zeehandelaar visited an upper-level bureaucrat at the Animal Health Division, Dr. R. E. Omohundro, who told him that twelve of the roan antelopes tested positive for the virus and four did not. This meant not only that the twelve antelopes would be denied entry into the United States, but also that the "entire shipment," including the four healthy animals, "would have to be destroyed." Moreover, the USDA would not allow the animals to be reexported because it wanted to limit their movement within the quarantine facility as much as possible to prevent the possible infection of a shipment of valuable Charolais cattle. Such contamination was possible, according to Zeehandelaar, because the understaffed facility used just one "boy" to serve multiple barns, meaning that viruses could be spread easily.[29]

To save the animals, Zeehandelaar went to downtown Washington to plead with the Agricultural Research Service's Dr. Mulhern. Mulhern, the 1967 recipient of the AWI's Schweitzer Award, perceived the seriousness of the situation and agreed to test the Charolais cattle to determine whether, in fact, they had been infected by the antelopes.[30] When the tests turned out negative, Zeehandelaar returned to New York to find someone who would ship and receive the reexported animals. Locating a taker required "time and diplomacy" because the animals had been refused by the USDA. After contacting several people overseas, he found a boat bound for Tunis that would also accept the original certificate of health from the government of South Africa, but the Tunisian government backed out of its initial promise to import the animals. In response, Zeehandelaar investigated "other possibilities, too many to enumerate." He went back to Washington, D.C., and while there, Tunis again agreed to take the animals. He found two boats bound for Tunis, but both were leaving after the deadline that the USDA had given him to remove the animals. Mulhern, while refusing any "indefinite extension," promised him a little more time. Finally, Zeehandelaar put the antelopes on a boat bound for Antwerp in late September, just days before the rumored start of an East Coast dock strike, and the animals were saved.[31] Such extraordinary efforts typified Zeehandelaar's personality, but their necessity also reflected the USDA's orientation toward the commercial animal industry.[32]

Even as Zeehandelaar and the AAZPA were working to improve the USDA's treatment of zoo animals, other interest groups were focusing attention on animal welfare problems within zoos. These groups perceived the real abuse and neglect of animals occurring not at USDA quarantine facilities, but inside zoos and other animal facilities, many of which escaped federal regulation. This animal welfare movement eschewed Zeehandelaar's personal, behind-the-scenes method of changing the system, preferring instead to publicize the animal abuses taking place in research laboratories and zoos, bringing congressional attention to the problem. By 1966, the movement could take credit for the passage of the Laboratory Animal Welfare Act, commonly called the Animal Welfare Act, which authorized the Secretary of Agriculture to regulate the transportation, sale, and handling of research animals. Soon thereafter, the activists took aim at other facilities, including zoos, that the 1966 act did not cover. For zoo directors who already felt that their interests were neglected by the USDA, the prospect of giving the agency even more oversight of zoo animals was truly frightening.

Federal Animal Welfare Legislation

By the late 1960s, the AAZPA began to recognize that it would have to both look out for its members' interests in the political system and push its members to serve their own interests by protecting the welfare of animals in their care. This dual defense was crucial to ensuring the future of public zoos, as illustrated by a teenager from Grand Junction, Colorado, who in 1969 wrote her local paper to offer the opinion "that the condition of the City Zoo should be brought to the attention of the public, and-or to those who are in a position from which they are able to do something about the situation." In 1970 her opinions reached some of those in the latter situation, when a House subcommittee considering amendments to the AWA read her letter. The congressmen heard about "starved, old, ugly animals sheltered within filthy dirty cages," including a lion who "looks like he's from the ice age, his hair is even falling out." The young author finished dramatically: "Grand Junction—the All American City! Well, the next time you visit the Zoo think about what it is to be an All American. Are you proud of what you see?"[33] For zoos to survive, they had to convince politicians and the public to answer with a "yes," but as the letter made clear, they had some work to do. Unfortunately, in 1970 the AAZPA lacked the political skills and resources to influence the proposed AWA amendments, faced a better-organized opponent in the animal welfare movement, and represented too many inferior zoos.

Although the zoo directors who ran large elite zoos were politically agitating for endangered species preservation internationally and trying to save animals from USDA indifference, there were many less progressive public and private zoos around the country. Indeed, as animal welfare groups were beginning to point out, numerous inferior zoos dotted the American landscape. As a result, attacks by animal welfare groups emerged in the late 1960s that put zoos on the defensive. The assault began with the Defenders when it embarked on a "crusade" against roadside zoos, uncovering and publicizing in their *Defenders of Wildlife News* a range of abuses, including inadequate space, filthy cages, isolation, and poor veterinary care.[34] Defenders stressed that only federal oversight would correct these abuses, because few state governments provided even minimum legal standards for humane care and those that did dedicated inadequate resources for their enforcement.[35] Another, and more publicly visible, critic of zoos by the decade's turn was the HSUS.

By 1970 the HSUS's mission had become infused with "moral [and] ethical concerns" and inspired by Joseph Wood Krutch's *The Great Chain of Life* (1957), in which he urged Americans to move closer to "an intimate relation with the natural world." For John Hoyt, the director of the HSUS, this meant that the organization should expand its animal saving work to include both wild and zoo animals. From its Washington, D.C., headquarters, the HSUS joined Defenders in scrutinizing zoos. Even though the HSUS identified with the goals of the larger environmentalist movement, its leaders felt that environmentalists neglected animal issues in favor of land and pollution concerns. Nevertheless, the HSUS and other animal protection groups eventually used some of the legislative successes of the environmentalist movement, such as the National Environmental Protection Act, as weapons in their legal battles against zoos.[36] By the latter 1960s and early 1970s, the HSUS and a range of additional animal protection groups, including the AWI, and the Society for the Prevention of Cruelty to Animals were arguing before the public that too many zoos, particularly the so-called roadside zoos (small commercial zoos), were inhumane institutions that should be severely regulated or abolished altogether. These groups represented formidable opponents for zoos because of the skills they had developed in a decade-long struggle to pass legislation regulating the treatment of laboratory animals.

Congress was not immune to publicity about cruel roadside zoos, and in 1969 Congressman G. William Whitehurst (R-VA) introduced five bills relating to animal welfare. Two of these had potentially enormous implications for zoos. The first would extend the AWA protections to all mammals and birds moved in interstate commerce or used by an institution receiving federal funds; the second called for the creation of federal agency to advance the welfare of zoo animals, largely through research and funding for improved zoo facilities.[37] Whitehurst, who became a tireless congressional advocate for endangered species preservation and the humane treatment of animals, had a personal interest in helping and reforming zoos because his wife, Jane Whitehurst, had revived the defunct Norfolk Zoological Society in the latter 1960s to improve the Norfolk Zoo.[38] By 1970, the Subcommittee on Livestock and Grains of the House Agriculture Committee held hearings on Whitehurst's amendment, and the AWI could report that a "spectacular change in climate of opinion" had taken place in Congress, with even the scientific community expressing reserved support.[39] The AAZPA was totally unprepared for this new political reality and missed several opportunities to present the zoo community's interests to Congress.

The zoo community remained silent during this public and congressional debate, the victim of inexperience and an inattentive parent organization, the National Recreation and Park Association (NRPA). Not an independent organization in 1970, the AAZPA received its official legislative services from the NRPA, which employed a legislative consultant in Washington, D.C., to keep track of park-related legislation and provide regular information about those legislative activities to the various NRPA branches. In an unofficial capacity, the National Zoo's Reed and Perry worked with government officials "on almost a daily basis" to represent the AAZPA's position on conservation and other zoo issues. President Ronald Reuther recognized that the AAZPA owed these two men "a debt of gratitude" for their political activity but would be better served by a full-time executive secretary. As he pointed out, Reed and Perry had a zoo to run and could hardly be expected to give their full time to tracking all zoo-related legislation.[40] By late November 1970, the AAZPA leadership began to realize that they were relatively blind to an animal welfare bill that was moving toward them at top speed.

Lester Fisher, the vice president of Chicago's Lincoln Park Zoo, sounded the alarm after attending a presentation at the annual council meeting of the Institute of Laboratory Animal Resources (ILAR). He came away with a "rather disturbed feeling," suddenly aware of the mounting pressure on government to regulate the procurement and welfare of zoo animals. After talking to government officials and ILAR members, he concluded that new regulations were "almost an academic question" and that animal welfare legislation would surely pass within the next year. Although not wanting to push the "Panic Button," Fisher noted that he knew of no attempt by the NRPA's legislative people to participate in the upcoming hearings on the AWA amendments.[41] Something was wrong, but it would be a few weeks before Fisher and others realized how wrong.

In fact, Perry was quite aware of the impact that the proposed AWA amendments would have on zoos, and he had made several efforts earlier in the year to get the NRPA to monitor the situation. In mid-March, he met with the NRPA's legislative consultant and asked for her assistance in tracking the progress of Whitehurst's amendments. In early April, he phoned her to check on the bills' status, and she assured him that "hearings would not be held this year." No stranger to the unpredictability of congressional action, Perry wrote her a few weeks later to reiterate the necessity of informing the AAZPA leadership immediately if any of the bills appeared "likely to reach the stage of hearings." In late July, Perry read with shock that hearings had been held on June 8 and 9 and that the Subcommittee on Livestock and Grains had listened to testimony by representatives from *every* major animal welfare organization as well as from laboratory research associations, but nothing from zoos.[42]

In the June 1970 House hearings on Whitehurst's amendments, much of the focus was on animals used for research purposes, but zoos emerged as a concern as well. Animal welfare groups argued for the inclusion of zoos by bringing forth the charges that groups such as the Defenders had made earlier in their publications. Christine Stevens, president of the AWI, noted that "charges of cruelty against both [*sic*] roadside, commercial zoos, and municipal zoos are common." To support her argument, she submitted a letter to the AWI requesting help for animals in a roadside zoo. The author, John Mehrtens, a zoologist and later the director of the Columbia (South Carolina) Zoo, noted his long-standing opposition to "the typical roadside zoo and exhibits of similar caliber" and had reported their "cruelties" to "various local humane societies" with "good results." Mehrtens described his recent visit to the Coxville Zoo, where he had found, among other terrible conditions, the "dessicated remains of an animal" in an unused cage and, in a 6 x 4 x 4 foot cage, a one-year-old Indian tiger so emaciated and crippled by rickets that it could not walk properly.[43] Mehrtens sought the assistance of the AWI, which he believed could adequately address the problem; he did not mention a need for government regulation. Ironically, Mehrtens's letter, along with other testimony, helped persuade legislators to broaden the focus of the amendment to animal exhibitors, including zoos of all kinds, not just roadside zoos. Newspaper editorials around the nation echoed the bill's sponsors in arguing that such protections were badly needed.[44]

The NRPA's legislative consultant had failed miserably to bring zoo voices into this debate, but the NRPA did little to prevent a repeat, and some within the AAZPA leadership did not immediately recognize the seriousness of the situation. When Fisher pushed the panic button in late November, it was already too late to stop or even modify the AWA amendments. A few weeks later, President Voss appointed a special committee to try to "block" the legislation, but as the chairman, Philip W. Ogilvie, pointed out, not only were his queries met with assurances that he was being an "alarmist," but in fact the legislation had already been passed when his committee was created. The AAZPA's NRPA-appointed executive secretary, Robert Artz explained with "embarrassment" that the House passed the bill on December 7 and the Senate did the same the next day without even sending the bill first to a Senate committee for hearings. President Johnson signed the Animal Welfare Act amendments (PL 91-579) into law two weeks later.[45] The amendments expanded the focus of the AWA from laboratory animals and pets to include exhibited animals. Zoos were now directly under the regulatory authority of the USDA through the Animal and Plant Health Inspection Service and the oversight authority of the House Committee on Agriculture.

The legislation promised to complicate the zoo business in numerous ways. First, it extended the 1966 definition of "animal" to include all warm-blooded mammals designated by the Secretary of Agriculture. More important, it directed the secretary to promulgate regulations for the humane care of these animals by zoos and other exhibitors. These regulations would cover "the transportation, purchase, sale, housing, care, handling, and treatment of animals" by zoos; the detailed nature of such regulations could be seen in the directive to establish minimum requirements for "housing, feeding, watering, sanitation, ventilation, shelter from extremes of weather and temperature, adequate veterinary care, and separation by species." In addition, zoos would have to become licensed by the USDA and keep "records with respect to the purchase, sale, transportation, identification, and previous ownership" of animals.[46] In light of zoo directors' prior experiences with the USDA, many reasonably dreaded the prospect of being so closely regulated by an agency for which zoos had never been a priority.

In fact, the USDA had strenuously resisted the AWA amendments. In its view, the responsibility of monitoring zoos and other exhibitors would siphon scarce resources from its real priority—agriculture. The 1970 amendments required the Secretary of Agriculture to establish and promulgate standards for the handling, care, treatment, and transportation of animals with which the USDA had little experience. Complaining about the difficulty of establishing and enforcing such standards for nonfarm animals, Agriculture suggested that "state and local agencies" enforce the AWA amendments and that the Department of Health, Education, and Welfare oversee the welfare of research animals because it was better versed in experimental design.[47] In general, the USDA argued that it lacked the technical expertise to evaluate the health and welfare of exotic animals. Congress, however, ignored the pleas to shift the jurisdiction of the act to other agencies, forcing zoos to develop a close working relationship with a bureaucracy charged with a mission it had not chosen.[48] The USDA's lack of knowledge about exotic animals had frustrated zoos in the past, but the earlier efforts by AAZPA leaders to forge open lines of communication with USDA officials proved useful in the months ahead.

Although some within the AAZPA initially believed that the AWA "would spell doom for many of [their] institutions" and even worried that the Department of Agriculture was "determined to close this nation's zoos," more experienced members recognized that the USDA would actually welcome assistance from zoos. As Perry noted, the new law was "exceedingly broad," and the USDA would be reluctant "to undertake an extensive administrative task." Nonetheless, he recognized that the animal welfare groups would use all means available to push for restrictive and additional regulations, and he recommended that his colleagues focus not on the formal rule-making process (as the NRPA recommended), but on the "informal" process. Only "ineffective" interest groups waited for rules notices of proposed rule making to appear in the *Federal Register*. If zoos hoped to influence the informal process that led to notices in the *Register,* they would need to become much better organized; otherwise, they would suffer another "fiasco," and the animal welfare groups would "write the regulations."[49]

Making up for the prior year's inaction, Oglivie and others immediately began the process of working with the USDA on the new regulations. Before the new year, Oglivie contacted Dr. C. O. Finch, the Agriculture official with final responsibility for writing the regulations, and by April several AAZPA representatives (including Zeehandelaar, Perry, and Reed) were invited to a meeting with nine high-ranking USDA officials to discuss a range of questions relating to the AWA. The meeting agenda and discussion that ensued made it clear that Agriculture desired significant guidance from the AAZPA on issues related to zoo animals. Following this formal meeting, AAZPA members met informally numerous times with individual politicians and USDA officials. A young legislative affairs committee member (and later chairman), Robert Wagner, emerged as key player in these discussions. Wagner had a keen insider's knowledge of zoos, boundless energy, and good political instincts about Washington, and he soon developed a friendly rapport with the USDA officials. Several years after the AWA's passage, Wagner recalled that the Department of Agriculture "sought [their] advice before implementing the Welfare Act and therefore, little, if any harm was wrought" on them.[50] On the other hand, a review of the AWA hearings indicated that, for much of the public, the line between roadside zoos and AAZPA zoos was not a clear one. Until the AAZPA could discover concrete ways to distinguish its members, every story about an animal suffering in a zoo threatened to implicate all zoos. The AWA heightened this danger, because animal welfare organizations would be able to use isolated examples of inhumane zoos to justify ever-stricter regulations. To avoid that situation, zoos would need to give much closer attention to both their political representation and their own animal care practices. In short, the AAZPA needed to raise its visibility and credibility as an organization dedicated to the interests of wild animals.

The AWA disaster revealed that for all of the conservation and political work performed during the 1960s by progressive zoo leaders and a few dedicated AAZPA committees, the association was only sporadically effective in promoting zoos' interests. To its credit, the AAZPA had joined international conservation efforts, it had supported legislation to restrict trade in endangered animals, it had lobbied the USDA to respect the unique value of imported zoo animals, and it had developed a system to police its own members' importation practices. Yet, after seemingly taking the lead nationally and internationally to address a conservation crisis and improve the welfare of captive animals, AAZPA leaders found themselves at the decade's end subject to a rising tide of federal legislation that did not make sharp distinctions between their zoos and roadside zoos. Although several key AAZPA members enjoyed relatively good working relationships with politicians and bureaucrats in Washington, D.C., the emergence of the animal welfare movement and the rapid legislative developments in the late 1960s indicated that the organization would need a more focused political strategy in the years ahead. Even before the passage of the AWA, the more active AAZPA members had concluded that the organization needed radical restructuring so that it could more effectively control its members, participate in the political process, and enhance its member zoos' professional credentials. To realize these goals, the AAZPA would need to separate from its parent organization, the NRPA, and the AWA fiasco proved to be the final piece of evidence that convinced the board to go forward as an independent organization.

Becoming an Independent Zoo Association

Founded in 1924, the AAZPA had existed as a branch of the American Institute of Park Executives until 1965, when it moved into the NRPA. In earlier decades, when many zoos perceived their mission as largely recreational, this relationship with public parks had made sense. By 1970, however, it had become obvious to the highly educated and self-consciously professional AAZPA leadership that their vision for zoos encompassed far more than recreation and that the distance between the NRPA's concerns and their own was significant. These leaders harbored a number of practical and philosophical criticisms of the NRPA, most of which revolved around their growing sense of professionalism and their dissatisfaction with the political representation provided by the NRPA. The NRPA struggled to hold onto the AAZPA, but by 1971 it broke away and for the first time assumed full responsibility for representing its members' interests.

Presidents Voss and Reuther led their colleagues in challenging the NRPA's lack of professionalism and its inadequate representation. These men perceived their relationship with the organization as "a relic of the old days when Zoos were nothing more than menageries." The scientifically and professionally oriented zoo directors found that the NRPA offered them little. The NRPA provided job placement services for zoos, but from the perspective of zoo directors, it lacked any ability to screen job candidates adequately. One assistant director from a major zoo described the candidates for a curator position provided by NRPA's professional placement services as "grossly unqualified" for *any* position in his zoo. This sense of dissatisfaction was heightened by the NRPA's decision in the spring of 1970 to fold the AAZPA's newsletter into a combined newsletter for all affiliates, a move that only the AAZPA voted against. Without editorial control, the AAZPA could no longer provide scientific information to its members, because the NRPA preferred general interest articles for its *Parks and Recreation* journal. The NRPA also decided that, beginning in 1972, the AAZPA annual conference would be merged with the parent organization's annual event. The lack of consultation with the AAZPA about such changes "angered" Reuther, but subsequent conversations between Conway, Reed, and NRPA officials failed to resolve the areas of disagreement, with Robert Artz characterizing the AAZPA as "unusually demanding and critical of small details," "nit-picking" even. A few months later, Artz had to admit that the NRPA's lack of attention to details was responsible for the AWA being debated and passed without any formal AAZPA input.[51] From that point, tensions between the two organizations grew worse.

Discussions after the passage of the AWA focused skepticism about the NRPA's ability to monitor the impact of such legislative developments on zoos. Voss, Reuther, Oglivie, and Perry argued strenuously that the incident revealed the inadequacy of NRPA representation and pointed to the necessity of the AAZPA assuming a firm legislative leadership role. These sentiments were echoed in a March 1971 AAZPA board meeting, where members vented about "short-comings in the [NRPA] service to AAZPA members," including "lack of attention to details, impersonal handling of membership affairs," slow "news distribution," and inadequate public affairs assistance. The legitimacy of their complaints seemed confirmed by the NRPA's legislative report for members later that month—not one of NRPA's "Public Affairs Priorities" directly affected zoos, nor did any of the described legislative initiatives.[52] Years later, Wagner recalled that the inappropriateness of continued affiliation with the NRPA was symbolized by the association's publication of an illustrated article that "heralded the killing of a tiger by U.S. troops while on R&R in Vietnam."[53] Thus, for both substantive and public relations reasons, AAZPA members voted at the 1971 annual conference to sever their relationship with the NRPA.

The work that lay ahead for the new president, Gary Clarke, and his colleagues was substantial. He, Wagner, Reed, Voss, Conway, and Zeehandelar all knew that it would not be enough just to separate from the NRPA. Implementing their various professional goals would not be easy. Without a robust accreditation program for their members, professional standards could not be enforced, nor would they would have much chance of preventing or at least influencing further governmental regulation. In a discussion about impending legislation such as the AWA, Voss, the legislative chair in 1969, reported that "legislation on the exhibition of wildlife cannot be stopped" and thus there was an "urgent need" for "AAZPA self-policing," He told the board that it needed to "(1) fabricate standards now and (2) move ahead as quickly as possible on zoo certification." Two years later Ogilvie reiterated this theme in the context of the AWA, arguing that the AAZPA's political goals "necessitate our establishing professional standards" for humane care.[54] Setting such standards before the USDA imposed its own would increase the likelihood that Agriculture would simply adopt AAZPA guidelines. The AAZPA would also need to sharply define its mission and values, partly through better publicity and partly through new membership requirements. Most challenging, the AAZPA would need to do all of this with an initial budget of around $50,000. Indeed, not until 1976 did the AAZPA have more than three employees, including the executive director.[55]

Central to the new AAZPA's image was its conservation mission. In its 1971 articles of incorporation, the AAZPA began by stating that it separated from the NRPA to "better pursue and further expand its involvement in conservation, science, and education." This conservation mission was reiterated six more times in the document. Its member zoos were described as "conservation agencies," its animal breeding and exchange policies were explained as having "conservation . . . and preservation purposes," and its central mission was stated as the advancement of "public education on the need for wildlife conservation and preservation" and the participation "in the international efforts of wildlife preservation." To reiterate the latter point, the articles concluded by stipulating that, in the event of the organization's demise, its assets would go to the IUCN "for the continuance of international programs concerned with wildlife conservation and preservation."[56]

The new AAZPA also clarified the requirements for and levels of membership, both individual and institutional. Perhaps recognizing their own central role in pushing other zoos toward a conservation agenda, the AAZPA's incorporators stipulated that only directors of zoos and aquariums could be "fellow members" or individual members. They also solidified the leadership role of large zoos by making institutional membership fees proportional to annual budgets. The larger the zoo, the more it paid, thereby providing larger zoos with an incentive to involve themselves more with the business of the interest group and giving more power to the kinds of institutions from which the AAZPA leadership was drawn.[57] Lower dues for smaller institutions would also give them an incentive to join by not burdening them with the same high fee imposed on large, prosperous zoos. They also conferred voting rights only to nonprofit institutions, seeking to distinguish AAZPA members from "roadside" and other commercial zoos that attracted negative attention from animal protection organizations. Specifically they allowed only "nonprofit, Federal and State tax-exempted, permanent-type establishments, open to and administered by the public."[58] This distrust of commercial zoos stemmed partly from the AAZPA's participation in the IUCN, which favored public, nonprofit, conservation-minded institutions over entities that used animals solely for economic gain. Similarly, Conway argued that the new AAZPA should exclude animal dealers from voting membership so that it would appear more "professional."[59] Nonetheless, for historical and pragmatic reasons, the AAZPA included a number of large, commercial zoos and aquariums among its nonvoting members, as well as an assortment of animal dealers and small, for-profit zoos. In 1971, it accepted any institution able to find a member sponsor and afford the annual dues.

On January 19, 1972, the AAZPA incorporated as an independent interest group, both to ward off increased legislation and regulation directed toward zoos and to further an agenda that included the conservation of wild animals and professionalization of the zoo business. From the beginning, these zoo directors were conservationists, working both internationally and nationally to preserve wild animals. They championed early species preservation legislation, and their zoos were recognized by some members of Congress as conservation institutions in the context of their time. Animal protection groups acknowledged that the directors of large city zoos such as Reed and Conway cared about preserving animals. They also knew, however, that not all zoos had such conservation-minded directors. As a result, both the AAZPA and the animal welfare groups recognized that after 1970 they would have to pay attention to smaller zoos and aquariums in cities around the United States. As we shall see, the AAZPA made it a priority to speed up accreditation and force smaller zoos to adopt the AAZPA's mission of conservation and education. For their part, animal protection groups tried to blur the lines between roadside zoos, small municipal zoos, and larger, scientifically oriented zoos.

TWO

On the Defensive

Creating a Strategy for Survival

In April of 1972 the American Association of Zoological Parks and Aquarium's import advisor, Fred Zeehandelaar, visited the Australian Embassy in Washington, D.C., accompanied by an official of the U.S. Department of the Interior and a representative of the Humane Society of the United States. To reach the meeting they passed through the lobby, "lavishly covered with kangaroo rugs." The purpose of the meeting was to discuss the fact that Australia allowed the killing of thousands of kangaroos but refused to export wild-caught kangaroos to American zoos. The Australian agricultural attaché explained that the kangaroos were damaging the crops of farmers, who were allowed to kill them to prevent further economic loss. Zeehandelaar suggested that Australians consider using the "democratic process" to introduce legislation that would at least allow these animals to be sent to zoos rather than be killed. As Zeehandelaar wrote in a follow-up letter to the attaché, "I would assume that a kangaroo would prefer to be incarcerated rather than be killed."[1]

Zeehandelaar's outrage about such cavalier treatment of wild animals echoed the growing concerns of the American public and its politicians about threats to wildlife. The fact that an animal dealer, a USDI official, and a humane organization representative could find common ground on the issue of protecting kangaroos illustrated that zoos' cooperation with government officials and animal welfare groups was possible. For zoos that had pressed for stronger government protection of animals, particularly from commercial interests, the political environment in the early 1970s was both welcome and foreboding. On the one hand, growing public interest in the environment represented a potential boon to zoos that promoted themselves as conservation centers. Bipartisan congressional support for action on the environment also promised to achieve meaningful protections of habitat and wildlife that many zoo professionals desired.[2] On the other hand, some of the new interest groups pushing for environmental legislation also implicated zoos in "crimes" against wildlife. Worse, the AAZPA lacked a clear political strategy for dealing with these groups and even for representing its members' interests before the federal government.

Two key pieces of legislation developed in this period—the Marine Mammal Protection Act (MMPA; 1972) and the Endangered Species Act (ESA; 1973)—provided much-needed protections for wild animals, but also threatened to prevent zoos from collecting or exhibiting certain wild animals. Zoos narrowly escaped these legislative prohibitions, but in its implementation of the ESA, Interior erected barriers of its own. Congress too kept the AAZPA on the defensive as it considered a number of proposals to create a federal agency to govern zoos. The pace of zoo-related government activity between 1972 and 1975—in 1973 alone, congressmen introduced nine bills affecting zoos—combined with the rapid proliferation of antizoo humane organizations, challenged the AAZPA leaders, who scrambled to develop a coherent and effective political response.[3]

In 1972 the AAZPA was a small interest group competing (and sometimes cooperating) in the political arena with a loose coalition of better-financed and more politically savvy animal welfare groups. The AAZPA had little money (its 1972 budget was $50,000) and depended upon the volunteer activity of just a few key members to advance its political and professional goals. Reviewing the lack of participation from individual members during his presidency in 1970, Ronald Reuther lamented the difficulty of fitting "able, *willing,* people . . . into positions of authority and responsibility, on about twenty-five committees."[4] Despite its inherent weaknesses, the AAZPA leadership developed a relatively coherent and generally successful political strategy to represent zoos' interests at a critical juncture in the development of U.S. environmental policy. Recognizing the precarious position of zoos in the debate about how to save wild animals, the AAZPA's legislative representatives worked hard to ensure that new laws and regulations provided zoos and aquariums with the limited right to "take" protected animals from the wild. Faced with the prospect of a federal zoo agency, they argued successfully for self-regulation. The AAZPA leadership had urged separation from the NRPA largely on the grounds of inadequate political representation, and after three years of independence they would be able to take some credit for taking the necessary steps to protect zoos' interests. These first few years of independence were not, however, without internal conflict, and by 1975 new complaints about the association's political leadership emerged.

Professional Standards and Professional Political Representation

As the Animal Welfare Act (AWA) debacle had illustrated, the AAZPA needed to pay closer attention to political developments in Washington, D.C. It also needed to immunize its members from charges that zoos were "inhumane." The latter task required that the organization intensify its efforts to professionalize the zoo industry by establishing standards of care and management for its member zoos. Because the existence of standards alone would not prevent further government regulation, the AAZPA also needed to develop a closer working relationship with the government agencies that affected zoos. Finally, the zoo community needed to reach out to the moderate animal welfare organizations and seek opportunities to cooperate with them. By 1973, the rough outline of this political strategy was largely in place.

The first part of the political strategy involved the creation and enforcement of professional standards. This initiative had begun in 1968, when the board developed a voluntary "registration" system to "upgrade the zoo and aquarium profession" by establishing recommended minimum levels of education for zoo personnel and developing continuing education programs. Even before the AWA took effect, the AAZPA's Education Committee took the step of surveying zoo directors about the existence of formal exams, rules, and instruction for their keepers. Chairman George Rabb needed this information so that his committee could "establish base lines [*sic*] and guides for the management of zoos worthy of respect." These initiatives took on new urgency once the inevitability of new government regulations under the AWA became clear. President Lester Fisher argued in 1973 that "the most important issue" facing the zoo community would be the "raising of standards of our physical plants, our keeper personnel and our Staffs." By the time they split from the NRPA, the AAZPA leadership realized that the association would have to exercise "meaningful controls" over member zoos to ensure that they met scientific, humane, educational, and conservation standards.[5]

Efforts during the 1960s to gain such control over member zoos' practices had been largely symbolic, but a first priority for the independent AAZPA would be the creation of a formal accreditation system for its member zoos. The kinds of standards sought by Fisher, Rabb, and others would help zoos demonstrate that they were meeting Interior and USDA requirements, and, if they became mandatory, they promised to weed out the weak zoos that attracted so much attention from the animal welfare groups. The accreditation program instituted in 1972, however, was only voluntary, a concession to the reality that many member institutions could not or would not meet the new professional guidelines. By 1980 the board had enough confidence in the AAZPA's value to zoos to make accreditation mandatory for new institutional members, but not until 1985 did accreditation become required for all members. In the meantime, the adoption of an accreditation system put the AAZPA on record as taking a proactive approach to improving zoos.

The rapid pace of government activity, however, indicated that, without effective participation in the political arena, any standards adopted by the AAZPA could be made obsolete or irrelevant by a new round of legislation or rules. Ideally, the AAZPA leadership wanted to shape the government rules that affected zoos rather than just react to those rules after they had been proposed or implemented. As John Perry, Theodore Reed, Reuther, and others noted during the AWA crisis, the zoo community was represented in Washington, D.C., only by volunteers. Lacking an executive director, a paid lobbyist, and even a permanent legislative committee, the AAZPA had no formal way to identify and pursue a political agenda. At the height of the AWA crisis (December 17, 1970), AAZPA president Gunter Voss partly addressed this problem by appointing Philip Ogilvie to chair a special legislative committee. In 1971, this committee was renewed, with Robert Wagner as the chair, and in 1973 it became a permanent standing committee. Thus began a period of experimentation with a range of voluntary and formal tactics to provide zoos with effective representation before Congress and regulatory agencies.

During its first full year of independence, the AAZPA relied on the legislative committee to track and respond to government activities. The committee prepared an annual report for members on new legislation and rules affecting zoos, and it provided legislative updates in the AAZPA's monthly *Newsletter*. Committee members testified at congressional hearings, met with government officials, and made recommendations to the AAZPA about positions to take. Presidents Gary Clarke and Fisher, as well as the AAZPA's first executive director, Margaret Dankworth—a longtime AAZPA employee who agreed to stay on for a few years—offered advice and encouragement to the committee during the first two years of independence, but their multiple responsibilities prevented them from assuming the kind of leadership role that some believed the association needed in this area. Zeehandelaar, for one, had long advocated the appointment of a full-time "Washington agent," but in the past, he claimed, his suggestions had met with silent disapproval from some board members. By 1973, however, the board, which now included Wagner, fully appreciated the necessity of securing professional political help, recognizing that "legislative strength must become a significant" part of how they would help their members. After reviewing several possible candidates, they agreed to hire William Hagen as the AAZPA's "legislative representative."[6]

Although not a "zoo man," Hagen had extensive experience in wildlife propagation and politics, having recently retired from a career in the U.S. Fish and Wildlife Service (FWS). A fisheries biologist specializing in hatcheries, Hagen joined the FWS in 1937. By 1950 he had been promoted to a Washington, D.C., desk job where he directed the Atlantic and Pacific salmon propagation programs. In 1962 Congress created the National Fisheries Center and Aquarium, a Washington, D.C.–based educational institution, and Hagen became its acting director. The AAZPA's 1973 president, William Braker, had served as a consultant to the center, so he naturally turned to Hagan when looking for legislative assistance.[7] Hagen's insider's familiarity with a government agency that regulated zoos' animal transfers and importations made him a valuable asset to the AAZPA, as did his interest in public education, conservation, and captive propagation.

Hagen, however, knew that his expertise alone could not help the AAZPA achieve its political goals, and he accepted the job offer with the understanding that his consulting work would be only the "first step" in a political process that would have to involve the board, the membership, and even local zoo supporters. First, the board would need to provide him with policy positions on issues such as interstate animal shipments and then clarify the extent to which he would be authorized to speak for that policy; second, the board would have to quickly decide a course of action against legislation that it opposed. Would it, for example, be "prepared to use every legal method to muster opposition forces?" Grassroots lobbying would require the AAZPA to mobilize support from other wildlife associations, from zoo directors, as well as from politically connected zoo trustees and board members. As Hagen put it, "Pressure is the name of the game," but success would depend on "how far you want to go."[8] Under Hagen's leadership, the zoo community began to participate more fully and systematically in the political process, but it still faced vexing problems regarding how to position itself against the increasingly vocal and well-funded animal welfare organizations.

Zoo leaders regarded animal welfare organizations as both potential allies and dangerous foes, and they disagreed about how to deal with them. Some of the organizations—such as the HSUS, the American Humane Association (AHA), and the World Wildlife Fund (WWF)—approached questions of animal welfare from a human-centered, scientific management perspective. They shared with zoos the belief that wild animal populations should be protected, but that protection did not exclude the possibility of killing or capturing some animals in a carefully managed manner. Moreover, because these groups did not object in principle to zoos, only to inhumane exhibits, they could prove useful in the AAZPA's efforts to help zoos secure funding to improve their facilities. The American Humane Association seemed particularly receptive to such a cooperative relationship, and in May 1972, with the support of the AAZPA, its "wildlife consultant," Richard Denney, surveyed AAZPA zoos regarding their economic needs; this survey report was then given to

Congressman G. William Whitehurst (R-VA) as supporting data for his bill (HR 6803) to create a National Zoological and Aquarium Corporation to assist zoos financially. For several years thereafter, AAZPA leaders complimented their AHA counterparts on their "educated and generous view of zoos," invited them to attend AAZPA annual conferences, and kept them informed of actions such as an AAZPA resolution condemning "rattlesnake roundups." For their part, AHA leaders pledged to continue their "joint efforts" on behalf of animals. In 1974 the AAZPA created a special liaison committee with the AHA.[9]

The AAZPA sought similarly cooperative relationships with other humane organizations, always seeking a benefit for zoos, but not always satisfied with the results. Individual zoos provided financial and rhetorical support to the WWF, a sister organization of the International Union for Conservation of Nature and Natural Resources (IUCN). When the WWF supported the successful efforts of Project Monitor—a coalition of protectionists that monitored MMPA permit activity—to block permits sought by animal dealers to capture 500 California sea lions for resale to the display industry, AAZPA president Reuther was not pleased. Although the WWF stood by its decision, Reuther's WWF counterpart offered "apologies" and reiterated that the WWF's position was "in no way anti-zoo." The HSUS also had frequent and generally friendly contact with the AAZPA through its wildlife representative, Sue Pressman, who had been the supervisor of animal health at Boston's Franklin Park Zoo in the late 1960s and knew many AAZPA leaders on a first-name basis. In her HSUS capacity, Pressman attended AAZPA conferences and lent support to "good" zoos, participating, for example, in a symposium on endangered and threatened species with William Conway, Reed, and Wayne King.[10] Like the WWF, however, the HSUS took positions that hurt zoos' interests, and the organization was viewed simultaneously as an ally and an adversary. In general, AAZPA leaders sought to educate humane organizations about "good" zoos, emphasize their organizations' shared interest in the protection of wildlife and the development of humane standards of care, and cultivate support for increased public funding for zoos.

Still, even as some AAZPA members developed friendly relations with animal welfare representatives, others worried that the points of shared interest were superficial and that it was naive to expect meaningful support from these groups. True, the AHA, HSUS, and WWF were generally supportive of zoos, but the animal welfare movement included new, more radical organizations that criticized the mainstream groups as being too conciliatory toward hunting and commercial animal interests. The Fund for Animals (FFA), Society for Animal Rights, Friends of Animals, Animal Protection Institute of America, United Action for Animals (UAA), Committee for Humane Legislation (CHL), and others opposed all hunting and commercial uses of animals and offered little or no support of zoos. Unlike the AHA and HSUS, which generally focused on abuses by roadside zoos, groups such as the FFA targeted zoos run by the AAZPA's most conservation-minded leaders, calling, for example, Reed's National Zoo "a virtual concentration camp for animals."[11]

The entry of these groups into national politics represented a real danger to zoos, one that they were not completely prepared to confront. In principle, the AAZPA favored legislative initiatives to fund zoos, protect marine mammals, and strengthen the Endangered Species Conservation Act (ESCA) of 1969, but the experience with the AWA had revealed that legislative details could be influenced by the animal protection groups. For the more pessimistic zoo leaders, the power of these groups appeared difficult to check. In 1972, the AAZPA's legislative representatives thus entered the political arena uncertain about their membership's political will and the dependability of their potential allies. The organization's political strategy made effective use of limited resources, but its success still depended to a large extent upon the volunteer labor of the legislative committee, and, because the president served only a one-year term, the political strategy lacked consistent leadership. These weaknesses did hamper the AAZPA's political effectiveness in the first few years of independence, but with each new crisis, the strategy evolved.

Protecting the Right to Capture and Import Endangered Animals

Prior to the 1970s, zoo animal importations had been regulated primarily by the USDA. As we have seen, zoos enjoyed a generally civil relationship with USDA officials and were able to work cooperatively with them on the implementation of the AWA. In 1972, however, Congress shifted its attention to endangered species. These new legislative initiatives both affected zoos' ability to acquire endangered animals and put them under the regulatory authority of relatively unfamiliar agencies. Among the independent AAZPA's first political challenges was ensuring that the MMPA of 1972 and the ESA of 1973 did not totally prohibit the importation of many animal species held by zoos and aquariums. Even while securing this protection, the AAZPA's political team struggled to develop a cooperative relationship with the Department of the Interior, which had been charged with enforcing the ESA. Personal attacks by zoo directors on Interior officials complicated the team's task and, more important, revealed that not all AAZPA members favored Hagen's "insider" approach to resolving political conflicts. By the end of 1973, dissatisfaction with AAZPA's handling of the MMPA and ESA led these members to form an independent zoo lobbying organization.

Public conservation education was written into the AAZPA's charter in 1971 and was used immediately to argue that zoos and aquariums should be exempted from the prohibitions against "taking" marine mammals and other endangered species being considered by Congress. Public enthusiasm for whales and other marine mammals fueled the development of massive outdoor aquarium parks, but it also generated intense political interest in restricting the killing and capture of those mammals. Christine Stevens, of the Society for Animal Protective Legislation (SAPL), and a few other activists capitalized on this public sentiment. They persuaded the Senate to pass unanimously a whale-hunting moratorium resolution in 1971 and then offered their support to the Ocean (later Marine) Mammal Protection Act proposed in the same year. Next, conservationists and animal protectionists pushed in 1972 for a more comprehensive ESA that would compel Interior to speed up its listing of endangered animals. Congress took up both issues, and, luckily for zoos, it proved open to the argument that their educational mission warranted special treatment under the law.

The MMPA was the first piece of domestic environmental legislation after the AWA that significantly affected zoos and aquariums. The MMPA offered protections to those marine animals not covered by the International Convention for the Regulation of Whaling. The original Senate version of this bill called for a nearly complete moratorium on the killing of all marine mammals, but opposition from the tuna and fur industries, as well as from mainstream conservation organizations such as the WWF and the Audubon Society, led Representative John Dingell (D-MI) to direct the drafting of an alternative, less-restrictive bill (HR 10420) that proposed to limit the taking of marine mammals through scientific management. Dingell, chairman of the Fisheries and Wildlife Conservation Subcommittee of the House Committee on Merchant Marine and Fisheries, held four days of hearings on both bills in September 1971. Statements by congressmen and government officials indicated the high level of support for quick action to save marine mammals, and the Animal Welfare Institute (AWI) declared the hearing the "most extensive . . . on animal protective legislation to date." The hearings also clarified which animal welfare groups were likely to support zoos in their desire to take small numbers of marine mammals for display purposes. The Friends of Animals, FFA, HSUS, CHL, SAPL, and Defenders of Wildlife backed the more restrictive Senate version, whereas the AHA, Audubon Society, and WWF favored the House version. When Dingell's committee reported the House version for a floor vote, it failed to pass, and in negotiations with the protectionists' lobbyists (led by Bernard Fensterwald), Dingell agreed to a compromise bill that passed easily in early 1972.[12]

Dingell's House hearings focused on the commercial killing of marine mammals, but testimony by Tom Garrett, representing Friends of the Earth, indicated that environmentalists supported the capture of marine mammals for zoos only when those institutions could provide conditions essentially equal to those enjoyed in the wild. Because both the House and Senate versions of the MMPA gave limited authority to either the Secretary of the Interior or the Secretary of Commerce to issue permits for the taking of marine mammals, AAZPA members, particularly those from aquariums, sought to influence the conditions under which such permits would be allowed.[13]

During 1972 Senate hearings on the MMPA, aquarium directors urged Congress to give aquariums the legal right to capture and display marine animals, arguing that as educational institutions they served the public interest by building awareness about and concern for endangered marine mammals. Even for-profit aquariums seized on their educational purpose during these hearings to justify why they should be included with "municipal and/or other nonprofit zoos" in provisions for capturing marine mammals. John Prescott (then general manager of Marineland of the Pacific, California) explained to Senator Fritz Hollings (D-SC) that "much of the public concern" about protecting marine mammals likely resulted from positive exposure to these animals at marine recreation parks. In fact, Prescott emphasized, 38,000 students visited his facility at no charge each year. A letter of support from Senator Alan Cranston (D-CA) stressed the "valuable educational service performed by these institutions." Senator Hollings himself reiterated this "public" service claim, concluding, "I am for the Marineland and the others."[14]

The argument that aquariums, even if run as for-profit institutions, performed a fundamental public service carried the day. The final version of the MMPA allowed permits for aquariums to capture marine mammals for educational display. The provision giving the Secretary of Commerce the right to grant taking permits represented a significant defeat for the more radical animal protectionist groups, which believed that Commerce would abuse its power. As if confirming this fear, within six months of the MMPA's passage, seven public hearings had been held on capture permits for aquariums, and the Secretary of Commerce had granted an "economic hardship" permit (which required no public hearing) to Sea World to capture eighty marine mammals for its Orlando, Florida, facility. More than half of these Sea World mammals died, thereby deepening animal welfare activists' suspicions about the motives and expertise of the aquarium industry in general and of Sea World in particular. The aquarium's image was not helped by its lobbyist, George Steele, who also represented the American Tunaboat Association and was perceived as responsible for a loophole in the MMPA that allowed the "incidental" killing of marine mammals by commercial fisheries.[15] Steele's subsequent association with the AAZPA deepened the rift between the zoo and animal protectionist communities and provoked internal debates among the AAZPA members about the wisdom of their political strategy.

Like the MMPA, the ESA emerged due to pressure from the animal welfare activists who felt frustrated by language in the 1969 ESCA that limited Interior to listing, and thereby protecting, only species that were in danger of worldwide extinction. The activists sought authority for Interior to place species on the protected list when they were "threatened" or "endangered," thereby increasing the chances that import restrictions and other efforts could lead to population recovery. Although initially reluctant to take on this mission, by 1972 pressure from the activists led Interior officials such as Assistant Secretary Nathanial Reed and the Office of Endangered Species' Earl Baysinger, as well as administration officials, to approach Congress about strengthening the ESCA. Again, Congressman Dingell played a lead role, directing his Fisheries and Wildlife Conservation counsel to draft a new ESA, shepherding the bill to its House passage in September of 1973 and then working out a compromise bill with the Senate just before the end of the year.[16]

The ESA gladdened zoo leaders, who had long argued that the government needed to do more to halt the commercial exploitation of wild animals and the destruction of their habitat. Equally important, the new act also guaranteed zoos' right to continue taking animals from the wild even as it limited the number that they could take and the ease with which they could collect them. From the beginning, many legislators and bureaucrats perceived the ESA as an effort to preserve animals in the service of human interests. In the 1973 Senate debates about the ESA, supportive congressmen returned repeatedly to the utilitarian value of nature, invoking Gifford Pinchot's so-called wise-use philosophy to justify the law's restrictions. For some senators, animals were a resource, like timber, to be preserved not for their own sake, but for the benefit of people. This resource conservation perspective meant that ESA supporters were not always firmly in the animal welfare camp. Senator Mark Hatfield (R-OR) saw no contradiction between protecting endangered species and returning them to population levels where they could be harvested for human purposes. As an example, he pointed to the research "value of primates to advance the biometrical and pharmaceutical sciences." Of more interest to the AAZPA, Senator John Tunney (D-CA) and others affirmed the aesthetic and recreational value of wild animals, noting that "we like to view them in zoos" and stressing that the ESA would encourage the captive propagation of endangered animals.[17] If zoos at times seemed to be protecting their own interests by capturing and breeding endangered species largely for public enjoyment, that is exactly what many congressmen intended when they voted for the ESA in 1973.

Congress clearly wanted to protect zoos' right to collect and exhibit wild animals, but in the same year that Congress debated and passed the ESA, a speech by the Assistant Secretary of the Interior, Nathaniel P. Reed, put zoos on notice that the agency charged with administering the ESA and other import-related laws harbored serious doubts about zoos' conservation credentials. On October 19, 1973, Reed gave a speech to an HSUS meeting in Atlanta, Georgia, that electrified the zoo community. He invited his listeners to join him in targeting "second-rate zoos" and the wild animal pet business to "solve some serious problems" related to the welfare of endangered species. Although he acknowledged the existence of "humane" zoos, Reed argued that there was "nothing humane or educational about animals caged like criminals" in the National Zoo in Washington, D.C., and others like it. Rather than present the "pathetic spectacle" of caged gorillas and their endangered kin, zoos should limit their collections to captive-bred animals and create real "environmental education centers" featuring films about animals. In fairness, a few zoo directors were making great strides in captive propagation, but far too many zoos were "better described as *consumers* of wildlife than as *conservers* of wildlife." Reed intended to use the power of the import permit system to change that situation. First, only zoos with demonstrated captive propagation programs would receive import permits; offspring from those zoos could then supply other zoos. Second, Interior would soon publish proposed rules to greatly expand the number of "injurious animals" whose importation could be denied under authorization of the Lacey Act. With the help of the HSUS and other humane groups, Reed promised to do battle with those opponents, including "some zoological societies," that would "work to maintain the status quo."[18]

The zoo community's response to this speech revealed its political inexperience and likely damaged its relationship with Interior over the next year or so. In late February of 1974, President Braker sent a copy of Reed's speech to all AAZPA members, inviting them (and their zoo society members) to respond directly to Reed while an official AAZPA position was being drafted. This open invitation subjected Reed to the intemperate suggestion that he and other Interior officials be "replaced at the USDI by their pictures" and the sarcastic observation that "maybe after I dig out from under the piles of Federal red tape I'll have time late this evening to do something for the animals in [my] zoo." Even Reed's "open letter" of explanation to the zoo community published in the September *AAZPA Newsletter* did little to mollify his harshest critics, who, rather than accept his offer of reconciliation, continued to criticize him. Former AAZPA president Clarke was among the few who seized the opportunity to "commend" Reed for opening "this channel of communication" and thank him for bringing some issues (e.g., the sale of surplus zoo animals) to the AAZPA's attention. Reed thanked Clarke for his civil remarks and confided that "the majority of the letters I have received from directors of reputable zoos have been so strident, vindictive, and inflammatory that continuing dialogue would be next to impossible."[19]

Few in the zoo community appeared to have considered the possibility that Reed's speech, however critical of zoos, was really intended to defuse some of the hostility directed at Interior by the animal welfare groups. The department had long supported the interests of ranchers, hunters, and other wildlife consumers. (Until 1974, the FWS was named the Bureau of Sport Fisheries and Wildlife.) This orientation accounted for Interior wildlife management programs that killed tens of thousands of wild animals a year to protect ranching interests from wild predators and other "nuisance" animals. Wolves, ferrets, birds of prey, and coyotes had been poisoned and hunted until some species were nearly extinct. Intense pressure from animal welfare activists, however, forced the department to take more seriously the goals of the 1960s endangered species laws and led its officials to regard any antihunting organization as a "10-foot giant armed to the teeth."[20] Now Interior desired allies in its mission to protect wild animals, but few zoo directors stepped forward, in part because the AAZPA leadership did not develop and implement a carefully considered response to Reed's speech.

The zoo community needed a good relationship with Interior, because Reed and other officials were developing a much tougher posture on the importation of wild animals. They helped draft the new ESA legislation, retooled the department to become more enforcement focused, and proposed new importation rules under the Lacey Act. Clark Bavin, the FWS's law enforcement chief, began recruiting agents from the FBI, the Bureau of Customs, police departments, and military intelligence. For his part, Reed kept his promise to give these agents new rules to enforce. On December 20, 1973, one day before the ESA became law, the Department of the Interior published proposed regulations that would expand the list of wildlife "injurious" to human beings, agricultural interests, forestry, or wildlife resources and thus subject to importation bans. Zoos would be exempt from this ban, but only after going through a new and complex import permit application process to justify their need for new animals.[21] As Reed had made clear in his speech to the HSUS, he planned to make sure that only zoos with successful education and captive propagation programs were granted permits.

Unhappy with the proposed injurious wildlife rules that appeared likely to make it difficult and even impossible for many zoos to import animals, Braker raised the possibility of hiring a lawyer to deal with Interior but first sent Wagner to talk with Earl Baysinger, Deputy Chief of the Office of Endangered Species and a key figure in the drafting and subsequent implementation of the ESA and the Convention of International Trade in Endangered Species of Wild Fauna and Flora. Former AAZPA president Fisher had previously enjoyed "friendly" conversations with Baysinger about pursuing a cooperative relationship between Interior and the AAZPA, and Baysinger had spoken at a 1973 AAZPA board meeting. In the spirit of those conservations, Baysinger assured Wagner that "USDI was not in pursuit of our zoos and aquariums and that [they] had little to fear regarding injurious wildlife." At the same time, he warned that the AAZPA should refrain from aggressive legal action against the USDI because he "had at his disposal a whole battery of attorneys 'sitting over there twiddling their thumbs and [he] would enjoy putting them to work.'"[22] In the short term, however, the AAZPA did retain legal counsel and prepared written comments on the proposed injurious wildlife rules.

In its brief, the AAZPA returned to the strategies that had proven so effective in Congress: stressing zoos' support for conservation legislation and pleading for their special status as uniquely powerful educational institutions. The AAZPA began by reviewing its decade-long involvement with the IUCN, its efforts to restrict members' animal imports, its lobbying for the ESCA, and its members' endangered species propagation programs. Most important, it stressed that, because zoos attracted more than 100 million visitors a year, they had an "unmatched opportunity to build understanding of wildlife problems" and thereby sustain public support for conservation legislation. Zoos were "urban national parks" where important public education on conservation issues took place. Therefore, the proposed injurious wildlife regulations would actually retard the conservation cause by limiting zoos' ability to give the public sympathetic "personal contact with and exposure to wild animals." Specifically, the AAZPA sought to exempt its member zoos—and *only* its member zoos—from the import permit application procedures, which it declared "unnecessarily cumbersome" and an "unfair burden." In this request, the AAZPA was careful to distinguish its "legitimate" zoos from those with no educational or conservation mission. In light of the fact that only 0.6 percent of the 120,000 animals imported in 1972 went to zoos, it seemed reasonable to the AAZPA to exclude zoos from these regulations.[23]

Wagner, Hagen, and others pressed their case at Interior's open hearings on the proposed regulations. Animal welfare organizations attended as well, and the Animal Welfare Institute claimed that opponents, including some zoo representatives, "behaved in a raucous manner seldom seen at government hearings, hissing representatives of conservation and humane groups who crossed the will of wildlife exploiters."[24] By contrast, Wagner believed that the AAZPA's representation at the meetings "was quite successful." Hagen predicted that final regulations would be published in 1975, and Wagner

Congressman G. William Whitehurst (R-VA) posing with Jethro, a Canadian timber wolf, in a news conference at the Capitol in 1973 to publicize his wolf-protection legislation. A tireless advocate for animals, he introduced numerous bills to regulate and support zoos. *Virginia Daily Press,* February 27, 1973. Courtesy of the *Virginia Daily Press.*

'I SHOULD TAKE THEM OUTSIDE THE TWELVE-MILE LIMIT AND DUMP THEM BECAUSE THEY MAY HAVE HOOF AND MOUTH DISEASE?'

This cartoon appeared in 1966 when the USDA demanded that zoos destroy all animals that might have hoof and mouth disease. *OLIPHANT © UNIVERSAL PRESS SYNDICATE.*

Lota, a forty-year-old Asian elephant at the Milwaukee Zoo in 1985. After showing aggression toward other elephants, she was given to a circus where she contracted tuberculosis. Her plight prompted celebrity protests and a lawsuit by Humane Society of the United States. *Milwaukee Journal Sentinel,* May 31, 2003. ©2005 Journal Sentinel, Inc., reproduced with permission.

Celebrity game show host and animal-rights activist Bob Barker joins a demonstration against Lota's transfer to the Hawthorn Company circus. *Milwaukee Journal,* January 6, 1991. Photo by Richard Wood © 2005 Journal Sentinel, Inc., reproduced with permission.

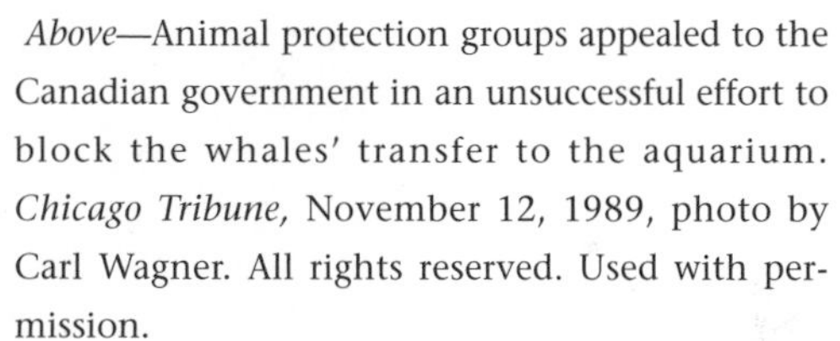

Above—Animal protection groups appealed to the Canadian government in an unsuccessful effort to block the whales' transfer to the aquarium. *Chicago Tribune,* November 12, 1989, photo by Carl Wagner. All rights reserved. Used with permission.

Left—Two beluga whales captured in Canada for exhibition at the John G. Shedd Aquarium, Chicago, swim in a temporary holding pen. *Chicago Tribune,* November 12, 1989, photo by Bob Fila. All rights reserved. Used with permission.

William Conway, longtime director of the Bronx Zoo and one of the key leaders of the AAZPA who advocated conservation and scientific research. Courtesy of the Smithsonian Institution Archives, Accession 96-024, box 23, William Conway.

Right—Timmy, a lowland gorilla played a role in the AAZPA's program to increase the captive population of endangered animals. Activists sued to prevent his transfer from the Cleveland Metroparks Zoo to the Bronx Zoo in 1991. He adapted well at the Bronx Zoo, however, fathering thirteen offspring before being moved to the Louisville (Kentucky) Zoo in 2004. D. DeMello © Wildlife Conservation Society.

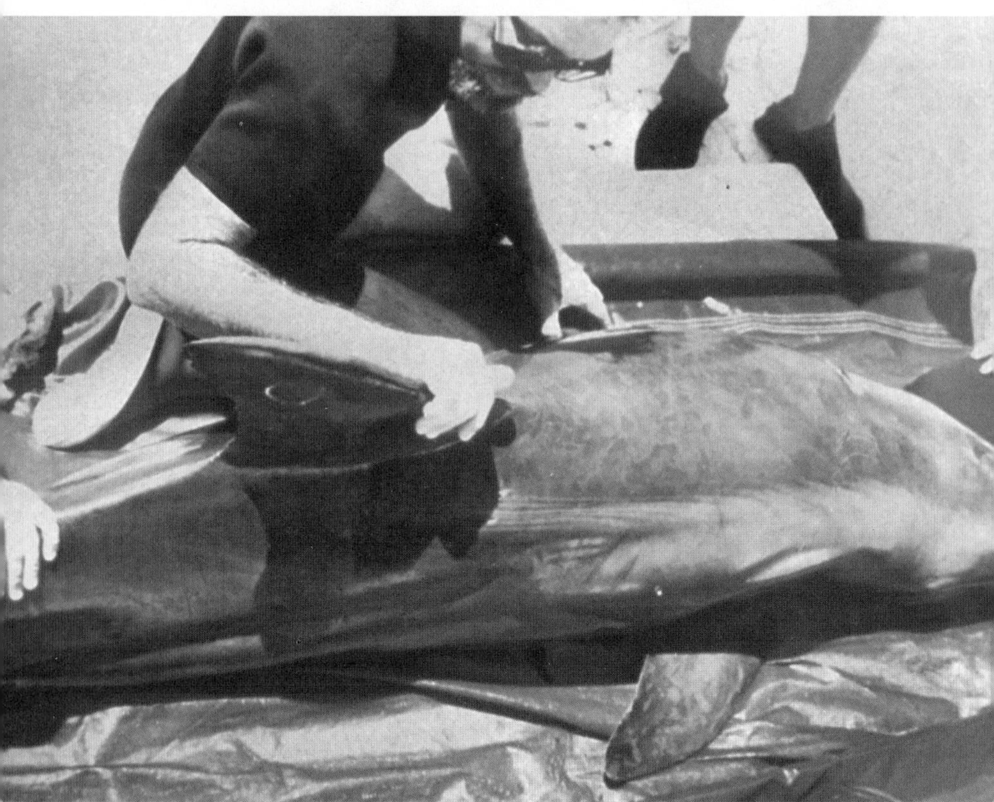

Buck, an Atlantic bottlenose dolphin, and one of his rescuers. Illegally released off the Florida Keys by two animal rights activists, Buck was suffering from hunger and an injury most likely caused by a ship propeller. The episode illustrated that animal protection organizations were sometimes better equipped to criticize zoos than to care for exotics themselves. Courtesy of the Dolphin Research Center, Inc., Grassy Key, Florida.

Buck looking much healthier at the Dolphin Research Center. Courtesy of the Dolphin Research Center, Inc., Grassy Key, Florida.

"KATIE" & "PERCIVAL" THE LIONS

Rescue story:

Zoos rarely have space available to accept unwanted or abandoned exotic pets. In 1993 The Detroit Zoo rescued two lions that were about the same age as two mother-reared lions here at the zoo. The female "Katie" had allegedly been used as a "guard dog" in a crack house. Her owner had become afraid of her as she grew, and a concerned citizen reported it to the Michigan Humane Society, where she spent several months before coming to the Detroit Zoo. "Katie" now socializes normally with both of the mother-reared lions, but has become intolerant of the rescued male "Percival", who was rescued in May of 1993, and so they are not kept together.

These two lions are the product of private breeders who sell cute little lion cubs to unsuspecting people who apparently don't realize that they've bought a very dangerous animal. The baby they're cuddling one day will quickly become a serious risk to themselves, their families, and their communities. They will have a very hard time finding another person or organization to relieve them of their purchase.

Unlike other cats, lions have very complex and long-term social relationships, and lions that are sold as "pets" are denied the chance to learn proper lion behavior from their mothers. These lions can be difficult, if not impossible, to integrate into proper social groups with other lions. This is one reason most zoos won't accept rescued lions.

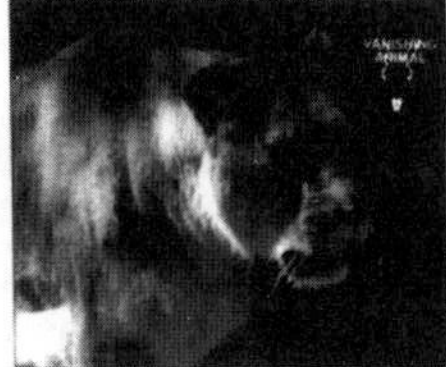

What can you do?

- Do not purchase exotic animals because they can be unpredictable, dangerous and difficult to care for.
- Choose an animal that you can adequately care for, and remember that maintaining a pet is a long-term commitment. Only domestic animals (which have been bred for thousands of years to live with humans) make good pets.
- Consider choosing a dog or cat from an animal shelter or humane society and have it neutered to prevent pet overpopulation.
- Encourage responsible local and national legislation to prevent this problem.

Sign at the Detroit Zoo. Like other AZA institutions, the Detroit Zoo opposes exotic pet ownership and has taken the additional step of becoming the home for dozens of animals rescued from abusive situations. The sign explains that Katie had been used as a "guard dog" for a crack house and warns visitors that lions do not make good pets. Authors' photograph, 2003. Courtesy of the Detroit Zoological Institute.

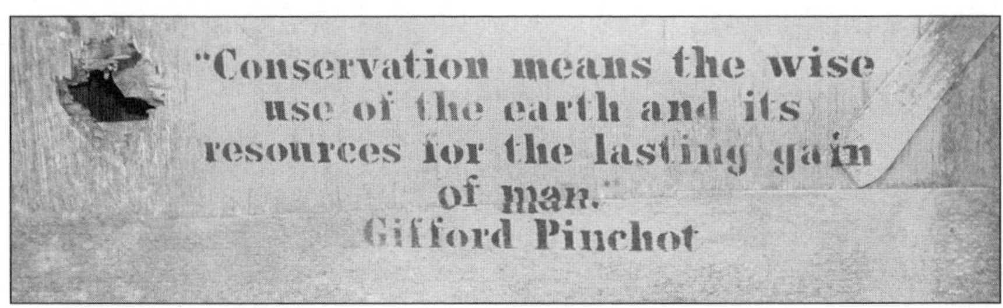

Zoos sometimes make their political ideology explicit, as the Columbus (Ohio) Zoo does here with a Gifford Pinchot quotation espousing a human-centered conservation philosophy. Authors' photograph. Courtesy of the Columbus Zoo and Aquarium.

reported optimistically that Interior appeared likely to ease zoos' burden somewhat by granting two-year blanket import permits rather than requiring a new permit for every transaction. In an apparent gesture of goodwill, Bavin attended the AAZPA's fall 1974 board meeting to explain the anticipated permitting process under the ESA and apologize for delays. Wagner later complained privately to the board about the lack of input and advice from the AAZPA membership during these negotiations with Interior, but the energetic politicking by him and a few other key zoo leaders continued.[25]

By mid-1975 the anticipated injurious wildlife regulations had still not been published, and the AAZPA turned to a sympathetic congressman for help. Robert Leggett (D-CA) arranged for a meeting between himself, the zoo community, several Interior officials, and a few animal welfare representatives. Wagner, Braker, and the AAZPA's legal counsel outlined the zoo community's concerns, particularly with the "high echelon" USDI officials responsible for this regulatory initiative. Leggett sided with the zoos, characterizing the proposed regulations as stretching the intent of the Lacey Act and promising to "discuss the entire matter at length with" Nathaniel Reed. Leggett's influence as the chairman of a subcommittee overseeing much Interior activity gave Wagner hope, but changes at Interior—President Gerald Ford was expected to nominate a new secretary, and several lower-level officials were replaced—continued to delay a final resolution. Then, the election of Jimmy Carter and further personnel changes prevented any decision until August 1976, when Interior announced that, instead of going forward with the proposed rules, it would ask Congress to act legislatively on the matter.[26] Although not the resolution that the AAZPA desired, zoos had at least been protected in the short term by the AAZPA's persistent lobbying on this issue.

From the outside it appeared as though zoos had fared quite well during this period of new endangered species legislation and regulations. Congress had affirmed their right to collect and import endangered animals, and Interior, with somewhat less enthusiasm, had backed off from proposed regulations that would have limited that right. Not all within the zoo community saw things quite so brightly. In October 1973, President Fisher informed the board that "the honeymoon euphoria was over," that "there are rump groups around and some disgruntlement with the Board over general issues and philosophies."[27] A faction of members, mostly from the larger commercial zoos, correctly recognized that the government's emphasis on the *public* educational value of zoos made them vulnerable. Aquarium directors, in particular, noted that the conservation activists who pushed for the MMPA objected strenuously to the public display of marine mammals and were likely to use the public comment periods to hold up aquariums' permit applications.

In September 1974, these concerned members met and formed an independent zoo lobbying organization, the Zoological Action Committee (ZooAct). The organization's officers came mostly from large for-profit zoos, as did its funding. No one from the AAZPA's traditional leadership ranks served on the initial board of officers, although Paul Chaffee and former president Ronald Blakely had joined it by 1976. Instead, its officers were sharp antigovernment types, such as Bob Truett, director of the Birmingham (Alabama) Zoo, who in a 1972 editorial about the AWA questioned "whether the bureaucratic monstrosity that is the USDA can do the job" of protecting captive animals and predicted that "in the hands of the USDA anything can go wrong and probably will." Headed by the professional lobbyist George Steele, ZooAct adopted an aggressive position toward the "humaniac" animal welfare groups and the government itself. Eschewing the AAZPA's moderate rhetoric, ZooAct members claimed that the MMPA and ESA were fundamentally "antizoo" laws pushed through by an extremist minority whose real goal was not conservation but the regulation and eventual abolishment of zoos.[28] This position reflected a nearly willful distortion of the history of both laws, but it started to seem more reasonable as zoos confronted radical animal protectionists in a new round of legislative activity related to the development of federal funding and control of zoos. By 1975 both the former and current AAZPA presidents, Braker and Blakely, were echoing ZooAct in the substance and tone of their communications with members, influenced in part by their experiences in 1974 with a series of legislative proposals to create a federal zoo agency.

The Promise and Threat of a Federal Zoo Agency

In the fall of 1974, Wagner recalled the past year as "extremely active, frustrating, and hectic," lamenting that, "before we can get organized to do battle" in one legislative area, another "pops up." Paralleling the injurious wildlife scare was a series of proposed bills that would have both provided federal funds to help zoos upgrade and established a federal regulatory agency to police zoos. Hopeful that a federal agency would infuse zoos with much-needed money, some AAZPA members also feared the control that might accompany those funds. As President Fisher put it, the zoo corporation bills were both "encouraging" and "frightening."[29] By the end of 1974, ZooAct members successfully rallied their colleagues to focus on the dangers represented by government regulators and animal welfare groups, and the AAZPA's political strategy shifted from cautiously optimistic cooperation with zoo-focused legislators to active resistance.

Congressman Whitehurst, who had played a significant role in the AWA and supported a range of wild animal proposals, had sponsored legislation to directly regulate zoos back in 1969, but not until 1973 did Congress give the idea serious consideration. In that year Whitehurst and Senator Mark Hatfield (R-OR) introduced "Federal Assistance for Zoos and Aquariums" bills (HR 1266 and S 2042) to create a National Zoological and Aquarium Corporation. Later in the year, the bills were rewritten with guidance from the AAZPA and reintroduced as HR 12047 and S 2774.[30] Hatfield was a middle-of-the-road conservationist whom President Nixon had previously asked to sponsor the ESA so Democrats would not have a lock on wildlife issues. Indeed, Hatfield's work on behalf of animals earned him the gratitude of the AWI, which later chose him to present its annual Albert Schweitzer Award for Outstanding Contribution to Animal Welfare.[31] Senate hearings on the zoo-assistance bills were held in January 1974, and for nearly two years the AAZPA provided input on this and a number of similar bills proposed by Congressman Dingell.

At the hearing for Hatfield's bill, the concerns of all interest groups became clear, as did the AAZPA's strategy for shaping legislation to its own advantage. The bill would have set up a government corporation controlled by a board with the authority to draft mandatory accrediting standards for zoos and aquariums. As a financial incentive, federal money in the form of direct grants and insured mortgages would help institutions construct new facilities to comply with the standards. The board could hire "additional personnel" as needed to carry out the act and provide technical assistance to institutions seeking accreditation. Although the outright grants of funds were only for any "public nonprofit zoo and/or aquarium," commercial zoos could also take advantage of the loans as long as they too were accredited.[32] Concerned largely with helping city zoos, the bill's supporters also hoped to entice roadside zoos into attaining accreditation and improving conditions for their animals. The bill's provisions were relatively favorable to zoos, in no small part because Wagner and Hagen had met in the fall of 1973 with Hatfield's and Whitehurst's legislative assistants to discuss modifications of the bill. To their surprise, they encountered "absolutely no resistance whatever": "both were receptive to [their] suggestions and comments," and the bills were rewritten "with a much more favorable attitude toward the AAZPA." Specifically, the revised bills provided for "outright grants rather than just loans" and modified the board composition to be somewhat more palatable to the zoo community.[33]

The zoo community turned out in force at the January 1974 hearings for Hatfield's bill, as did their opponents. Eight zoo directors, three zoo society representatives, and the AAZPA president made oral or written statements; Wagner and Hagen attended the hearing as well, though neither testified. Because the AAZPA had substantially influenced the nature of the bill, zoo witnesses supported it, particularly its promise of federal funding and its spirit of philosophical support for zoos. The AAZPA was concerned, however, about the details of accreditation and the board's composition. In his letter to the subcommittee, AAZPA president Braker noted that, "after several years of study of the standards and procedures for the accreditation of United States zoos and aquariums, including review of such programs already established here and in other countries," the AAZPA's accreditation program "would appear to be most efficient, and accomplish the purpose of the Bill in part." He suggested coordinating the "features of the Bill with the on-going accreditation program of the Association, thus avoiding undesirable duplication of effort." Former president Clarke reviewed the AAZPA accreditation process and suggested that it "be incorporated" as the accreditation program for the proposed board to avoid "duplication," "delay in implementing such a program," and "the expense of study and development necessary in establishing accreditation."[34] Clarke correctly perceived that, however useful to zoos, the proposed zoo corporation could replace the AAZPA's accreditation standards with new ones over which the humane groups would have some formal control. In such a system, zoos might find themselves subject to stricter standards.

Seeking to limit the influence of humane organizations, AAZPA witnesses argued for the inclusion of additional pro-zoo representatives on the board. As proposed, the board was to have seventeen members including the director of the National Zoo as an ex-officio member, the comptroller general, and fifteen other members appointed by the president; the latter fifteen would include two officers or employees of the Department of Agriculture; one officer or employee each from the Department of State, the National Oceanic and Atmospheric Administration, and the Department of the Interior; two professors of zoological park management; four representatives from national humane associations; and four zoo directors. The nongovernmental members would be appointed for six years and could serve only two terms. The board's composition was designed to balance the interests of the zoos and their critics by having equal numbers of each. Presumably the two professors who taught zoological park management would serve as impartial voices.

In response, however, both the AAZPA and the animal welfare groups lobbied for greater representation. AAZPA president Braker suggested the addition of a "zoological or aquarium society member" because these public-spirited citizens significantly assisted zoos. The CHL and the FFA, in contrast, perceived the board as already "stacked" in zoos' favor. Bernard Fensterwald, counsel to the CHL, wanted the act amended to "put some sort of balance" into the board. As drafted, he argued, "at least eight of the 17 members are zoo oriented, and only two [*sic*] are representatives of 'national humane associations,'" filling a role akin to that of "'token blacks' in so many organizations."

Stephen Seater, the FFA's field director, said that it would be "imprudent to establish an agency, comprised predominantly of zoo professionals, to regulate and make grants to zoological parks and aquariums." He, too, proposed that there should be "more representation on the Board from the humane organizations."[35]

Although the animal welfare activists were generally against the bill, their proposed amendments focused on requiring larger and more humane exhibit spaces for accreditation. The FFA argued, for example, that the board should "upgrade zoos" by "rejecting the menagerie concept of housing animals" and by funding exhibit renovations to increase the "amount of space for animals." Moreover, the group insisted, the board should include animal behavior specialists who knew how to "construct enclosures that allow animals to live in a natural condition where their behavior patterns can find expression." Similarly, Stevens, from the Society for Animal Protective Legislation, argued against any expansion of zoos "except in the sense of increased amounts of space allotted to animals." To help illustrate the need for this provision, she told the audience that in 1973 the Los Angeles Zoo had stuffed Galapagos tortoises "into a storeroom from October to May, where there was such an enormous infestation of rodents that they were bitten," leading to one tortoise's death.[36]

Taking a more radical position, the CHL wanted to abolish some zoos entirely and feared that the bill would have the unintended effect of supporting zoos that should be phased out, not propped up with temporary but ultimately ineffective improvements. In testimony before the subcommittee, Fensterwald described zoos and aquariums as "outmoded, inhumane institutions, which year by year, serve a less important function in our society." Most distressingly, he argued, this bill "specifically fosters the perpetuation of the worst animal prisons, the so-called roadside zoos" by allowing them to apply for financial support.[37] These "outdated" and "truly disgraceful" institutions "should not be given cosmetic aid at taxpayer's expense," but eliminated altogether. Echoing this desire to see zoos eventually disappear, the Society for Animal Protective Legislation wanted clear language that the bill would not fund increases "in either the number of animals or the number of zoos." Fensterwald followed his testimony with a *Washington Post* opinion piece in which he characterized the zoo corporation bill as a "bail-out program" for institutions that it was "time to phase out." Like Nathaniel Reed, he questioned zoos' educational value (favoring films instead) but went further and derided the idea that the captive propagation of endangered species constituted meaningful conservation. Hinting at forthcoming battles, he opined that the new ESA did not protect zoos' right to breed endangered species.[38]

Fensterwald also theorized that the zoo corporation bills had a sinister, hidden purpose. He claimed that biomedical researchers who wanted grants from the National Institutes of Health were exerting "tremendous pressure" to "turn the zoo into a research and breeding center." These researchers, Fensterwald argued, were "not satisfied with inflicting disease and suffering on millions of domestic animals in pharmaceutical and medical laboratories"; they now hoped to "obtain exotic species for various basic research experiments." He cited a National Society for Medical Research newsletter alerting its members that the AAZPA's 1973 annual conference would encourage the use of "zoos for biological and biomedical research." Fensterwald overstated the relationship between zoos and medical researchers, but some zoos did sell surplus animals to the research industry and conducted experiments on zoo animals. In addition, the Institute of Laboratory Animal Resources of the National Research Council had provided support for the AAZPA's 1973 conference, a fact that had disturbed several members and led Executive Secretary Margaret Dankworth to caution against inviting Congressmen Whitehurst and Dingell to participate in the conference. For his part, Whitehurst's interest in animal welfare was "pure"—he sponsored numerous bills to protect wild animals from hunting and trapping, for example—and he had reacted with dismay to the "misinterpretation of 'research'" being spread by radical groups such as UAA.[39]

Senate subcommittee chairman Claiborne Pell (D-RI) expressed sympathy for some CHL arguments but flatly rejected its more radical positions. First, he noted common ground on the issue of improving animal exhibits, reminding Fensterwald that the "purpose behind the bill is to recognize the inhumanity that exists" in many zoos. Pell, however, attributed the deplorable conditions to a lack of money, not to the actions of "local zoo societies and zoo directors." It is interesting to note that Pell's most extended criticism was of the CHL's objection to using zoo animals for research. He argued that if there were only fifty whooping cranes left in the world, he would use them for medical research to cure cancer. He later returned to his theme, getting Reuther, director of the Philadelphia Zoo, to agree that tuberculosis research conducted at his zoo had saved "hundreds of thousands if not millions of people's lives."[40] Pell essentially rejected the revisions proposed by animal protection advocates.

Senator Hatfield's bill never made it to a floor vote, but the experience before Pell's subcommittee surely shocked the zoo community. Anticipating horror stories about abusive roadside zoos, AAZPA members came prepared with detailed information about the value of the behavioral research projects of "good" zoos, the number of visitors they educated, the number of species they protected, and their promising future (if given adequate support). Hatfield's bill potentially insulated the professional zoos from the kinds of criticisms directed at roadside zoos, both by giving them the money to address the problem of outdated exhibits and by legitimizing their activities with the seal of federal approval. But prior relationships with the AHA and the HSUS, both of which supported the zoo corporation bills in principle, had not exposed zoo leaders to the force and fury of antizoo activists such as Fensterwald. After the hearings, the benefits of government financing appeared less significant than the costs of government rules that could be influenced by these unfriendly groups.

Congressman Whitehurst continued to introduce legislation designed to support zoos' efforts to create more humane zoos, but the zoo community showed only polite interest in these bills, most of which never had committee hearings. In 1974, he sponsored legislation to provide funds for zoos and aquariums to create "appropriate homes" for animals "no longer able to exist elsewhere." He envisioned the establishment of "survival centers" like the one that the National Zoological Park planned at Front Royal, Virginia, to assure the future of not merely of animal species, but zoos themselves. Pointing to forces that wished to see zoos eliminated, Congressman Whitehurst quoted Fensterwald, who wrote that "a growing number of people want zoos phased out by attrition" and believed that as "animals die off they should not be replaced."[41] Later in the same year, Whitehurst put forward HR 13004 to create a federal "Federal Zoological and Aquarium Board" to further regulate and support zoos, a proposal that he introduced every congress until 1985. Whitehurst's proposed federal board would have set "voluntary standards for accreditation for zoos and aquariums, and provide[d] funds for assisting these facilities in bringing themselves up to the desired standards."[42] He argued that, although the AWA had improved zoos, the government needed a board that would "raise their standards higher." Because zoos housed species that had become nearly extinct in the wild, the federal government had a responsibility to make sure that zoos provided the best homes possible. For the most part, the zoo community cautiously regarded Whitehurst as a well-meaning friend, but in 1974 he became associated with a "zoo control" bill that was distinctly unpalatable.[43]

Whitehurst's HR 1154 ("Federal Zoological and Aquarium Board") might not have gone far by itself, but the more-powerful Congressman Dingell sponsored a similar bill, HR 16458, and by late 1974 Dingell was rewriting the proposals into HR 70, which he and Whitehurst introduced in 1975. Their bill would establish a "Federal Zoo Accreditation Board in order to insure that zoos and other animal display facilities maintain minimum standards of care for animal inventories," and would provide "technical and financial assistance to zoos" for renovations and breeding.[44] Dingell had been instrumental in the drafting and passage of the MMPA and the ESA, and he wanted to use zoos as a resource for endangered animals. He had concerns, however, about the safety and trafficking of zoo animals that he hoped HR 70 would correct.[45]

Within months of their testimony in favor of Senator Hatfield's bill, zoo directors were working feverishly to obstruct HR 70, which they referred to as the "zoo control bill." Although AAZPA members' testimony before Congress had been courteous and positive in January 1974, their subsequent private communications about Dingell's bill were less so. In his September 1974 legislative report, Wagner updated the board of directors on Dingell's bill, which he noted was "prepared with assistance from Christine Stevens, of the Society for Animal Protective Legislation," and the HSUS's Pressman, a veterinarian who investigated and publicized inhumane conditions in zoos. Dingell had "sought no input from the AAZPA." Wagner believed that the bill was "very dangerous" because it proposed regulating space requirements, mandating personnel standards, and requiring accreditation. Further, it allowed for citizens suits, which would lead to "public harassment."[46] Similarly, a 1975 *AAZPA Newsletter* bluntly informed zoo professionals that Dingell's "H.R. 70 would essentially, regulate *every* zoo and aquarium." One of the penalties for lack of federal accreditation, the newsletter warned, was that zoos would not be able to "engage in animal commerce," the very lifeblood of zoos prior to the widespread practice of breeding programs.

The accreditation threat was quite real, because only a handful of zoos had applied for AAZPA accreditation by 1975 and presumably a federal system would accredit similar numbers. In 1974 the AAZPA had urged the federal government to adopt its own accreditation standards, but, unlike Hatfield's bill emphasizing financial benefits for accredited zoos, Dingell's bill imposed restrictions on unaccredited zoos and offered fewer benefits. The provisions requiring that "zoos and other animal display facilities maintain minimum standards of care for animal inventories" were seen as especially galling, because the AWA gave that authority to the USDA, which had recently initiated several meetings with zoo professionals to determine minimum cage sizes for exhibiting animals. For zoo directors already complaining about the "virtually uncontrolled proliferation of bureaucratic regulations and laws which threaten the continued existence" of zoos, the prospect of an additional regulatory agency was intolerable, an unjustified attempt by "Big Brother" to either "socialize" zoos or "put them out of business completely."[47] In letter and spirit, HR 70 appeared more closely aligned with the interests of zoo critics than with those of zoos. AAZPA members were not entirely in agreement, however, about how best to respond to Dingell's bill. Conservation Committee chair King argued against "condemn[ing] the whole bill" and joined Wagner in recommending that the AAZPA try to work with Dingell to improve it. But, whereas Whitehurst invited the AAZPA to help him draft HR 1154, Dingell ignored AAZPA offers to assist on HR 16458 and HR 70. Rebuffed by Dingell, President Blakely turned to grass-roots lobbying, urging AAZPA members and their local zoological societies to write their legislators. By May, the House Subcommittee on Fisheries and Wildlife Conservation and the Environment to which HR 70 had been assigned emerged with an even "more restrictive" Federal Zoological Control Board bill (HR 6631). Continuing their behind-the-scenes lobbying, Wagner and Hagen met with the subcommittee chairman Leggett to discuss their concerns.[48]

In contrast to the quiet approach taken by King and Wagner, ZooAct, under the guidance of Executive Director Steele, launched a public relations campaign to keep HR 70 from even receiving a hearing. Adopting alarmist rhetoric, ZooAct characterized government regulation of zoos as counterproductive to efforts to save endangered animals. The title page for a trifold ZooAct brochure read, "Exotic Birds Slaughtered by USDA! USDA Impedes Protection of Endangered Species! Zoos Prohibited from Breeding Endangered Species! Marine Mammal Act Stymies Scientific Research! Growing Federalization Will Make Today's Zoos Extinct!" Inside, the brochure warned that the proposed Federal Zoo Accreditation Board would "regulate literally every aspect of a zoo." An accompanying press release characterized this regulation as removing zoos from the sensible oversight of local government and putting them "at the mercy" of federal bureaucrats. ZooAct's secretary/treasurer, John Mehrtens, director of South Carolina's Columbia Zoological Park (now called the Riverbanks), wrote opinion pieces and gave interviews in which he portrayed the USDI as simultaneously obstructing zoos' endangered species propagation programs and "*developing* endangered species" with its Federal Predator Control Program.[49] In all of these public relations activities, ZooAct appealed to those skeptical of big government, an approach that found favor with President Blakely, who praised ZooAct's activities in the *Newsletter.*

Powerful congressmen shared ZooAct's and the AAZPA's distaste for the creation of another federal agency. Working on behalf of ZooAct, Mehrtens enlisted the aid of his district representative, Floyd Spence (R), his antigovernment regulation senator, Strom Thurmond (R), and his other senator, Ernest Hollings (D). In statements read into the *Congressional Record* in 1974, all three congressmen expressed their opposition to the federal regulation of zoos. Hollings, for example, warned that "humane societies would dominate pressure on this agency." All three urged their fellow congressmen to read journalist John Chamberlain's recent *Chicago Tribune* interview with

Mehrtens. The article made the case that a federal zoo agency was a bad idea because the Department of the Interior had failed to prove itself a competent steward of animals, causing, for example, the poisoning deaths of coyotes, foxes, and even pet dogs. The article also questioned the federal government's fiscal commitment to zoos, worrying that "in a period of inflation" it "would hardly be willing to provide money to make zoos better or build up their breeding banks of endangered species." Citing the advice of Mehrtens, Chamberlain proposed that, "rather than have a timorous and poorly funded Washington bureaucracy running our zoos . . . and doing the usual sloppy federal job of it," the "American Association of Zoological Parks and Aquariums should take responsibility for the animal show much as doctors and lawyers provide professional competence in their own ranks."[50] The combination of lobbying strategies employed by ZooAct and the AAZPA prevented HR 70 from receiving a hearing, and, though zoo regulatory advocates continued to introduce legislation throughout the 1970s, no federal zoo board bill ever reached a floor vote.

Despite its satisfactory outcome, this round of legislative activity heightened the zoo community's awareness of their critics' power. AAZPA leaders emerged more convinced than ever that they needed to speed up their accreditation process if they hoped to prevent increasing governmental involvement in zoos' work. The internal discussions about antizoo groups, however, had also revealed divisions among members and dissatisfaction with the AAZPA's leadership. Beginning with the formation of ZooAct, a faction of largely commercial zoo directors began to pursue its own political strategy. The AAZPA had survived its first few years of independence, but, if it hoped to continue to lead the zoo community, it would have to appease the critics within its own ranks.

Putting on a Public Face

The AAZPA entered the 1970s arguing that zoos served a unique public education function. Within five years, however, the AAZPA was quietly debating how best to gain public approval for its member zoos without subjecting them to complete public control. When Wagner studied Congressman Dingell's Federal Zoo Accreditation Board bill in 1974, he found its "most serious sections" were those "dealing with civil and criminal penalties and *citizen participation*."[51] Wagner feared a situation where various antizoo citizen groups gained the legislative right to participate in zoo policy making. And yet, the AAZPA limited its *voting* membership to professionals from public and nonprofit zoos and aquariums, institutions which citizens already had some ability to monitor through their local elected officials. The AAZPA leaders who emerged in the 1960s were philosophically committed to the idea that zoos served a public interest and that interest was largely conservation and education focused. The NRPA had been a bad fit for these leaders, precisely because they did not perceive zoos' mission to be primarily recreational. The paradoxical elitism of these public-minded AAZPA leaders, however, pitted them against the growing ranks of commercial zoo and aquarium directors who wished to be more fully involved in the AAZPA and who were more vulnerable to animal welfare groups' claims that they were "exploiting" animals. The AAZPA thus needed to find a way to address the commercial members' needs and concerns without damaging its image as an organization representing "public" zoos.

This preference for public zoos had more to do with the AAZPA's leaders' perception of commercial zoos than with a commitment to transparent zoo operations. For-profit, commercial zoos, in the view of many AAZPA leaders, too often were unprofessional, "one-man operations of the roadside variety" that neglected their animals' welfare. In general, commercial zoos were potential liabilities for the zoo community because they attracted unwanted criticism from animal protection advocates and sullied the reputations of "good" zoos. On the other hand, some of the larger, more professional commercial zoos (e.g., Sea World, Busch Gardens) promised to be important allies in protecting zoos from citizen complaints. Like many interest groups, the AAZPA had to figure out a way to keep these two potentially antagonistic constituencies—public zoos and private zoos—under its umbrella. In the long run, the political effectiveness of zoos would depend upon their ability to present themselves as a cohesive, unified group.[52]

The legislative climate added urgency to making an accommodation with the commercial members. As the creation of ZooAct indicated, commercial aquarium directors felt that the AAZPA responded too timidly to animal protectionists and government bureaucrats. Recognizing the growing forces allied against zoos, President Braker in 1974 recommended admitting "highly qualified" commercial zoos as full members. In an admission of motives, Braker pointed out that these commercial zoos could "better afford to contribute to a 'war chest'" for the fight against zoo foes. Unless the bylaws were changed, top-quality commercial zoos would not be eligible for AAZPA accreditation and would therefore be denied the "benefits" of accreditation that might someday be written into zoo-related laws and regulations. In a worst-case scenario, the AAZPA might face a rival organization of commercial zoos that would divide the zoo community's otherwise united voice. The trick would be to "bring the legitimate private groups into the fold, without opening the gates to a flood of roadside operations." In addition, the AAZPA would need to limit the power of its commercial members to protect its image as an organization representing "public" zoos.[53] Braker's advice, which the AZZPA eventually followed, turned out to be prescient in two ways. First, the AAZPA became financially stronger over the next few decades, partly because of the higher dues paid by commercial member institutions. Second, the AAZPA's strategy of portraying zoos as "public" institutions protected zoos from a range of legal challenges that could hardly have been anticipated in 1974.

In the short term, however, the experiences with the MMPA, the ESA, and the federal zoo agency bills had exposed some related differences of opinion between public and commercial zoo and aquarium directors regarding the AAZPA's political leadership. Broadly speaking, the commercial and smaller zoos faced off against the more established nonprofit zoos from which the AAZPA's leaders had been drawn for the past decade or so. Directors from the former group complained about the "ivory tower" mindset of the latter. In an open letter to AAZPA members in February 1975, one ZooAct member and several of his colleagues chastised both the AAZPA leadership and its general membership for responding too timidly to the antizoo forces. In their view, smaller zoos lacked the resources to comply with many new regulations, let alone keep track of and respond to proposed laws and regulations. For the AAZPA to become a truly representative and effective professional organization, more of its members would have to become actively involved in the political process and dues would have to be increased to fund political activity.[54]

These criticisms reflected a philosophical difference of opinion about how best to negotiate the political landscape. The established leadership favored a twofold approach: first, using the AAZPA to transform member institutions into professional, conservation-focused zoos that would be relatively immune from criticism; and, second, working quietly within the political system to cultivate good relationships with the government officials, politicians, and even animal welfare groups who took an interest in environmental policy. By contrast, the ZooAct faction preferred an aggressive campaign that put foes of zoos on the defensive. ZooAct literature exposed the alleged incompetence and ulterior motives of the antizoo forces, characterizing them variously as "radical," "extreme," "humaniacs," "little old ladies in tennis shoes," and "protectionists for profit."

ZooAct's aggressive tactics may well have complemented the AAZPA's more moderate approach by serving to rally the uninformed and uncommitted among the zoo community, but it also indicated that the AAZPA's political strategy was still a work in progress. The AAZPA leaders had helped the organization mature as a political actor by establishing a permanent legislative committee, securing Hagen as a legislative liaison, and coordinating sustained engagement with politicians and government officials. Nevertheless, ZooAct members raised valid points about the AAZPA's shoestring budget for political activity, the need for long-term leadership from an executive director, and the lack of widespread political participation from the membership. Although Hagen had been "tireless" in his "behind the scenes" efforts on behalf of the AAZPA, Wagner noted in 1974 that few members understood and appreciated his work, and fewer still provided advice when Hagen or Wagner requested it. The AAZPA had been moderately successful in the political arena during its first few years of independence, but Wagner understood that legislative activity affecting zoos would continue, and he worried that "the very life of the zoological park and aquarium profession is at stake."[55] The creation of ZooAct also revealed that the AAZPA had not fully addressed its members' frustrations with the emerging antizoo forces. The tension between these two AAZPA factions would not disappear over the next five years, but under the guidance of two AAZPA presidents who shared ZooAct's perspective—Braker (1974) and Blakely (1975)—the AAZPA moved toward a more aggressive political strategy to rally its members and protect its interests.

THREE

A Stronger Zoo Community

Legislative and Regulatory Victories

American Association of Zoological Parks and Aquariums president Ronald Blakely began 1975 by alerting his fellow zoo advocates to the challenges ahead: a "flood of federal regulations and legislation" and a host of well-organized, "radical" animal welfare groups. Zoos had underestimated the zealousness with which animal welfare activists were pursuing their cause through Congress, mistakenly banking on what they perceived as the peripheral nature of their opponents. Motivated by his fear of antizoo sentiment and legislation, Blakely used the bully pulpit of his *AAZPA Newsletter* each month to educate AAZPA members about the dangers they faced and urge them to "save zoos from extinction" by becoming politically savvy. He asked members to reach out to moderate animal-related interest groups and implored them to lobby when zoo-related legislation came before Congress.[1]

In his monthly letters, Blakely employed a rhetoric bordering on the hysterical, leaving AAZPA members with no doubts as to the serious consequences of their political apathy. Outlining the true sources of animal suffering, Blakely pointed to the "villains" of "overpopulation, environmental destruction, and the rape of nature by greed," reasons "too nebulous [for animal welfare activists] to form a good focal point" and "unify their members." For AAZPA members who might be inclined to sympathize with external organizations trying to reform zoos, he reminded them of their role in the "stringent, unworkable, and thoroughly unacceptable" Dingell "zoo control" bill. Not one to coddle the enemy, Blakely nonetheless wrote an "open letter to the [Humane Society of the United States] HSUS," in which he suggested that the two groups work together through the "AAZPA Humane Organization Liaison Committee" to eliminate "bad situations where they exist in zoos, humane groups or anywhere else."[2] In Blakely's view, the political landscape in 1975 was decidedly hostile to zoos, and only an invigorated, aggressive, and well-organized AAZPA membership could hold back the tide of radical animal protectionist groups.

To protect themselves from further federal controls, members of the zoo community needed to take more guidance from the AAZPA. In the 1960s, the AAZPA had attempted to get its members to adopt voluntary restrictions on the import of endangered animals, and in 1975 it faced the more urgent task of convincing its member zoos to seek accreditation. Likewise, it had to cultivate a more politically active membership. Finally, as an organization, the AAZPA needed simultaneously to work for its members' interests and continue to nurture its image as an advocate for animals, particularly endangered species. In the period between 1975 and 1980, the AAZPA's leaders did in fact make progress in these areas: they solidified the AAZPA's conservation image by supporting reauthorizations of the Endangered Species Act (ESA), demonstrated zoos' desire for stronger animal welfare regulations by supporting increased protections for animals in transit, developed a closer working relationship with the federal agencies that regulated animal exhibitors, and assisted AAZPA members in locating federal funds to support improvements in animal welfare and public education. Despite these successes, not all members agreed that the organization had struck the proper balance between supporting conservation initiatives and protecting zoos' interests.

Organizational Changes

After several years of independence, the AAZPA leadership took some necessary steps to strengthen the organization, particularly in terms of its political leadership and its control of members. The AAZPA's political representation became increasingly professionalized and centrally guided, in part because of pressure from commercial members and in part because of the leadership role assumed by Robert Wagner, who became the organization's first executive director in mid-1975. Wagner also emerged as a vocal advocate for the AAZPA's accreditation program, which promised to give zoos greater credibility when they claimed to be humane conservation institutions.

Blakely and his immediate predecessor, William Braker, pushed the AAZPA toward a more aggressive political strategy. During his presidency, Braker retained legal counsel to represent the AAZPA during discussions with Interior about injurious wildlife regulations, and when his aquarium colleagues established ZooAct in August 1974, he immediately formed an ad hoc committee to liaison with the ZooAct people. This committee—composed of Wagner, Blakely, and John Werler (the president-elect for 1976)—went through a "courtship" period in which it tested the amount and quality of cooperation it received from ZooAct. In May 1975 Wagner reported that, through March, ZooAct had provided "almost nil" input to him but that, in April, a "flurry of activity" showed that ZooAct could be "simply great in advising and assisting AAZPA." Impressed by George Steele's lobbying experience, Wagner concluded that a cooperative relationship would be "highly beneficial to both" organizations. President Blakely agreed, and in December 1975 the AAZPA entered into a formal contract with Steele, thereby assuring seamless communication between ZooAct and the AAZPA leadership. Steele's services would be paid for from a new Special Education and Information Fund (SEIF), separate from the AAZPA's general fund.[3]

While Braker and Blakely's embracement of ZooAct's "broadened, more aggressive response" to government and protectionist groups represented a shift in the AAZPA's political strategy, the organization made other, equally significant changes. First, thanks to a new membership dues structure, the AAZPA enjoyed greater financial stability by 1975. Its budget had increased from $50,000 in 1972 to $135,000.[4] Second, Wagner was made executive director in May 1975. In this position, he provided the organization with stable and active political leadership for the next seventeen years. Tutored by William Hagen and then Steele, Wagner was the kind of leader that Fred Zeehandelaar had long claimed the organization required: a genuine "zoo man" with political experience. His symbolic ascendancy as the organization's head could be seen in the fact that the *Newsletter's* cover page, previously reserved for messages from the president, was now generally Wagner's official forum. In Wagner the organization had someone who could not only assure continuity in policy, but also initiate and moderate discussions on important issues, including professional standards and accreditation.

Wagner accepted the necessity of a partnership with ZooAct, but he followed Reed and others in believing that an effective long-term political strategy included taking steps to force the improvement of member zoos. As we have seen, since the 1960s a handful of progressive zoo directors had worked through the AAZPA to both improve zoos' practices and pressure the government to protect endangered animals. These directors came from zoos that groups such as the Animal Welfare Institute (AWI) had acknowledged as serving "educational and scientific" purposes.[5] When Congressman G. William Whitehurst, the HSUS, the AWI, and others spoke about the deplorable conditions in zoos, they referred mainly to the commercial roadside zoos. Unfortunately, not all AAZPA member zoos in 1975, even public ones, gave sufficient resources to providing humane living conditions for their animals and a meaningful educational experience to their visitors. In fact, AAZPA leaders such as Ronald Reuther, Philip Ogilvie of the Portland, Oregon, Washington Park Zoo; and Gary Clarke of the Topeka Zoo (Kansas) wrote articles in the early

1970s pointing out that public zoos too often were only "glorified amusement parks" that paid "minimal attention to the conservation, well-being, comfort, interpretation, and scientific study of their animals." Size did not assure quality, and animal welfare groups occasionally singled out large public zoos such as those in San Francisco and Los Angeles as requiring reform.[6] Until these and the AAZPA's many smaller public zoos adopted standards and practices similar to those of the best zoos, the entire American public zoo community ran the risk of being characterized by its worst members.

A major stumbling block in the AAZPA's drive to improve the image of zoos was its members' failure to understand the importance of voluntary accreditation. The AAZPA leadership had recognized the political protection afforded by accreditation in the late 1960s, when they created a "voluntary registration plan" designed to "upgrade" their profession. Facing the prospect of Dingell's "zoo control" bill in 1974, Wagner recommended that "the Accreditation Program must somehow be speeded up to relieve some of the pressures confronted by the Board and the Legislative Committee in discussions with governmental agencies."[7] To earn credibility with congressmen and bureaucrats, the AAZPA needed to police its member zoos. If the AAZPA could rapidly increase the number of accredited member institutions, it would be in a better position to argue that further laws were unnecessary and to play a role in shaping the inevitable regulations stemming from the Animal Welfare Act (AWA), the ESA, and the Marine Mammal Protection Act (MMPA).

Although the AAZPA had an accreditation program in place by 1975, in reality it exercised little control over its members' animal care practices because so few members had even applied for accreditation. By 1975 it had accredited only *three* institutions from a pool of roughly two hundred potential member zoos.[8] Without accreditation, AAZPA membership by itself signified little in terms of an institution's quality or philosophical outlook, because institutional members included animal dealers, game-hunting ranch owners, and small commercial zoos. At this point in their political history, most public zoos were still small city institutions that saw little value in changing their policies and procedures to receive accreditation from a national organization that could confer few tangible benefits. These zoos complained directly to Wagner that accreditation hardly seemed worth the hassle of application. Changing this attitude became a priority under Wagner's leadership, and increasing the number of accredited zoos became the benchmark by which improvement in the profession's image could be measured. Identifying the lack of understanding about the regulatory threat from Congress, one member of the AAZPA accreditation committee complained that "most people" in the profession "do not understand the benefits of Accreditation or its process."[9] The AAZPA leadership understood quite well that individually

American zoos were vulnerable to attacks from animal welfare organizations but that by developing a stronger collective identity, zoos could survive and even prosper in an environment of closer federal scrutiny.

Accreditation alone, however, would not guarantee that politicians fully considered zoos' interests. To become an effective lobbying force, zoo professionals at the grassroots level needed to make their views known. Within the AAZPA a small core of zoo professionals from major institutions had long been comfortable networking far beyond their home cities, but many of their colleagues worked in small, intensely local environments. This latter group of AAZPA members was not always proficient in communicating its policy preferences to elected officials at the national level. There were exceptions, of course. Directors in Albany, Georgia, and Tulsa, Oklahoma, convinced their local governing commissions to pass resolutions opposing Dingell's Federal Zoo Accreditation Board bill on the grounds that "the operation of zoological parks and zoos is not a matter involving federal problems, but is a local matter."[10]

Unfortunately, member enthusiasm for national politics was not widespread or well informed, particularly in the arena of rule making. In one typical example, the AAZPA analyzed members' responses to a 1976 Department of the Interior rule making on nonhuman primates, one that would obviously affect most member zoos. Wagner complained that the AAZPA had sent copies of the proposed rules to 695 members, but only 14 of them had actually contacted the Department of the Interior. If the AAZPA hoped to become a political force, Wagner chided, it would need "much more membership response to legislative matters." At times the AAZPA also chastised its members for their lack of political acumen. Hagen asked members to consult with the AAZPA office "regarding Federal regulations before contacting the Federal agencies or Congressional delegations," because "premature protests often deteriorate our relations with various regulatory authorities." Seeking to address members' political inexperience, in 1977 Steele proposed holding a political action workshop at each of the AAZPA's annual conferences. The board endorsed this proposal, recognizing the value of having "a few key people around the country 'trained' to organize a grass roots campaign." The annual workshop led eventually to a stand-alone legislative conference held in Washington, D.C., so that attendees could meet their congressional representatives face-to-face.[11]

Wagner's leadership role in these endeavors cannot be overemphasized. Because government action affecting zoos tended to move slowly, the AAZPA benefited from the long-term perspective that Wagner brought. He gave the AAZPA a public face, a single person with whom zoo directors, government officials, and animal welfare leaders could interact. By bringing Steele aboard, Wagner assured that the zoo community's political interests were monitored full time by an experienced lobbyist. No longer would important pieces of legislation such as the AWA and the ESA be debated in Congress without AAZPA input. At the same time, by educating the general membership about impending legislation and rules, Wagner sought grassroots support for the AAZPA's political positions. Finally, by stressing the necessity of accreditation and embracing the adoption of a mandatory "code of ethics" (1976), Wagner completed the transformation of the AAZPA into a professional organization, one that could assure policy makers that its member zoos conformed to a series of guidelines that defined them as humane, conservation institutions.

The ESA and Conservation in the Zoo

In July 1974, the Great Adventure safari park opened in New Jersey. Great Adventure had been issued permits by the Office of Endangered Species (OES) to import twenty-four tigers on the basis that on-site breeding would "enhance the propagation of the species." After six of the tigers were killed in fights, animal welfare groups and the media descended, leading to federal investigations of the facility and the threat of a civil suit by the OES.[12] The episode illustrates why animal welfare and conservation groups looked skeptically at zoos' claims that they were "protecting" endangered animals by breeding them. This episode also helps us understand why the AAZPA wanted to limit its membership to institutions willing to meet its accreditation standards, which increasingly included strict guidelines about the breeding and exchange of endangered animals. If AAZPA member zoos hoped to convince skeptics that their practices and mission were substantially different from those of institutions such as the Great Adventure animal park, they required a professionally managed breeding program, combined with public support for the conservation goals embedded in federal legislation. From 1975 to 1980, the AAZPA labored on both of these fronts. Guided by Wagner, the AAZPA's political strategy regarding the ESA and Interior became increasingly more sophisticated by involving greater numbers of members, coordinating the zoo community's participation, and creating alliances with conservation organizations.

Unlike Great Adventure, the AAZPA wanted to limit the number of animals taken from the wild. By breeding animals already in captivity, zoos could both protect wild populations and preserve species for display. Zoos could successfully propagate many captive animals, and advances in veterinarian medicine promised to extend this capability to additional endangered species. The AAZPA imagined the creation of a large "captive self-sustaining" population of endangered animals that would not only supply zoos but also provide specimens for reintroduction to the wild.

The zoo community believed, with good reason, that Congress had intended for the ESA of 1973 to support zoos' captive breeding programs. In addition to provisions specifically allowing import permits for zoos engaged in captive propagation, Senate floor debate indicated that the bill's supporters imagined that one of its central purposes would be to stimulate the recovery of endangered species through captive breeding. Both the bill's sponsor, Senator Harrison Williams (D-NJ), and its floor manager, John Tunney (D-CA), stressed "propagation" as a central purpose, mentioning zoos several times in this regard. Senator Pete Domenici (R-NM) likewise believed that the ESA would stimulate efforts to "renew and refurbish deteriorated species." In language that implicitly supported the idea of zoos as arks, Domenici argued that "various programs of captive propagation would be beneficial for rare and endangered species in order that progeny raised in captivity could be used to replenish the wildlife population."[13] In keeping with their general dedication to the conservation of endangered species as resources to be enjoyed by people, these senators clearly expected that the ESA would motivate Interior to take proactive steps to increase the stocks of endangered animals.

Soon after this Senate debate, but before the ESA became law, the AAZPA board of directors considered the bill's implications. The USDI's Earl Baysinger emphasized to the board that "education" would no longer be "a determining circumstance to allow importation of endangered species" as it had under the Lacey Act and the ESCA of 1969. In anticipation of Interior's new permitting criteria, the AAZPA board gave the Conservation of Wildlife Committee permanent status and revised its mission to focus on the expansion of captive breeding programs for endangered species.[14] But, to prevent genetic inbreeding and maintain an adequate supply of animals for display, zoos needed to be able to exchange individual endangered animals freely. Such exchanges among zoos, however, became complicated by the Department of the Interior's reluctance to make permitting rules distinguishing between the importation of wild-caught endangered animals and the exchange of captive-born endangered animals.

When the Department of the Interior listed a species as "endangered," captive specimens suddenly came under the same restrictions as wild ones, meaning that even the loan of an animal from one zoo to another required an ESA permit. By 1975 the AAZPA had secured temporary permission for its zoos to obtain permits for breeding loans of endangered animals, but it was still waiting for a ruling that would allow "freer movement" of captive-born animals. Unfortunately, Interior made little progress in developing rules or processing endangered species applications, a situation the AAZPA attributed sarcastically to a "lack of personnel (or typewriters)." Wagner saw the problem as one of "red tape," and he hoped that AAZPA testimony before a House subcommittee oversight hearing on the ESA in October 1975 would create the necessary political pressure to speed up Interior's pace. From the AAZPA's perspective, "the efforts of zoos and aquariums to propagate endangered species must be encouraged and assisted by USDI—not thwarted and delayed." The AAZPA desired Interior to downgrade some animals' ESA status from "endangered" to "captive self-sustaining" and allow their exchange between zoos without permits. In late September 1975, just a week before the House oversight hearings, Interior did publish regulations establishing a relaxed permitting process for interstate trade in captive self-sustaining species, but zoos continued to fear that captive-born populations would decline unless animals could be moved more quickly to take advantage of breeding opportunities.[15]

The AAZPA and its member zoos engaged in some of their most intense lobbying to date to bring about rule changes on this issue. At the 1975 ESA oversight hearings, Wagner ensured that the nine AAZPA witnesses (including ZooAct representatives) presented a unified message. Hagen had suggested a general outline of key points to stress, and the AAZPA's legal counsel "reviewed every word" of each person's testimony. Steele testified first, setting the tone, establishing the zoo community's central concerns, and appealing shrewdly to key committee members. Steele began by bemoaning the "bureaucratic nightmare" and "thicket of red tape" that threatened to "put zoos out of business" and lamenting the "many extremists, both in and outside of government," who had perverted the ESA's intention by, for example, refusing to delist the American alligator, a species of commercial interest to Congressman John Breaux's (D-LA) constituents. Steele recommended that zoos be granted blanket exemptions for interstate trade in captive born animals and that commercial zoos be given the same status as public ones. Chairman Robert Leggett (D-CA) agreed with Steele on all points, establishing the committee's receptiveness to the remaining AAZPA witnesses. E. (Kika) De La Garza (D-TX) shared his frustrations with trying to help his local Gladys Porter Zoo (an AAZPA member) "wrestle" with federal rules, and James Oberstar (D-MN) endorsed Braker's suggestion that Interior simplify the permitting process by certifying AAZPA accredited zoos as eligible for permit-free exchanges of endangered species. The zoo community's concerns dominated the three-day hearings, and Wagner happily concluded that Chairman Leggett appeared sympathetic to the AAZPA's position.[16]

In response to pressure from the AAZPA and Leggett, Interior did loosen the requirements a bit so that zoos could obtain expedited permits for individual transfers that would "contribute to the breeding of other animals" by, for example, freeing up space, but not until 1977 did it establish a system by which zoos could apply for a blanket two-year permit that would allow them to exchange any animals from an approved captive self-sustaining population species with other permit holders.[17] The AAZPA reported dryly that, "in the opinion" of the Fish and Wildlife Service (FWS), this was a "simplified procedure," but in testimony before a 1977 House oversight hearing on the ESA, Wagner revealed that the permitting requirements for captive-born animals were "more stringent" than those for wild animals. As an example, he cited a San Diego Zoo application that took forty hours and seventy-five pages to complete.[18]

Even after zoos had completed the lengthy application, understaffing and funding problems in the FWS significantly delayed the approval process. By 1978, the OES had a 437-permit backlog. Dissatisfied with Interior's actions, the AAZPA pressed hard for its members to provide comments at a 1978 hearing on proposed captive-born rules and received the largest member response ever—seventy comments—though even this number was far short of the several hundred it desired. The lengthy battle to acquire captive-bred exceptions created much distrust, and when Congress's 1978 reauthorization of the ESA provided a special exemption for captive-born raptors the AAZPA doubtfully hoped that "Interior, in all its wisdom," would now "adopt a similar philosophy" for all other captive-born zoo animals. Eventually, in 1979, Interior began to place additional species on its exempted list, and the AAZPA tried to mend fences by praising the Secretary of the Interior for the "foresightedness" of his agency's decision.[19] By 1981, Interior listed forty bird and mammal species as captive self-sustaining.

Interior's eventual change of policy on the captive-bred wildlife issue owed much to the AAZPA's strong efforts to create scientific breeding programs that would distinguish its zoos from such outfits as Great Adventure. The long struggle with Interior clearly indicated that zoos required a rigorous, centrally controlled breeding program that could ensure the establishment of a reliable long-term supply of endangered animals for zoos, satisfy the government's concerns about traffic in endangered species, and even create surplus populations for reintroduction programs. In the late 1970s the AAZPA board began to discuss the creation of a master breeding plan, a centralized system of controlling breeding among member zoos. In 1979 it announced the planned implementation of this system, now called Species Survival Plans (SSPs), and explained how its "scientific and cooperative" nature would preserve endangered species for display purposes[20] Key to the SSPs was the control of breeding by "specialists" responsible for each listed species; the committee developing the SSP concept emphasized that management and breeding decisions by SSP coordinators should be considered "binding" on all zoos that held the affected species.[21] The emphasis on scientific management and binding participation by members was intended not only to ensure that captive populations were genetically healthy, but also to convince

Interior that AAZPA zoos could regulate captive self-sustaining endangered species populations without intensive government oversight.

In a less obvious way, the AAZPA's captive breeding programs also supported the FWS at a time when it was politically vulnerable. Not coincidentally, Interior changed its position on captive wildlife permits during the height of anti-ESA congressional hearings and just as the AAZPA implemented its SSPs. This change of heart reflected a growing perception on the part of FWS officials that the AAZPA could actually be an ally in its efforts to defend its record on the ESA. In December 1978, the chair of the AAZPA's wildlife conservation management committee, Dennis Meritt, attended a four-day midwestern fish and wildlife conference at which numerous state and six federal wildlife service officials were in attendance. Meritt was welcomed and even participated in a four-hour closed meeting on endangered species with wildlife service officials. Of particular concern to these officials were the "effects of outside interests on departmental policies." These officials felt that they were losing control of policy and feared that an upcoming congressional "review of USDI/FWS operations and budgets [would] profoundly affect the Department." They recognized that the AAZPA's endangered species breeding programs, if tied to USDI reintroduction programs, could deflect some of the criticism that the agency was doing too little to actually save endangered animals. In conversations with Meritt, these officials openly made "a plea [for] help" from the AAZPA. They requested that Meritt "act as a 'sounding board' and 'information passer'" to the AAZPA board in return for a "'listening ear' by FWS personnel."[22]

In light of the AAZPA's frustration with FWS policy during the mid-1970s, Meritt was understandably enthusiastic about this new attitude. The scientific research presented at the conference stimulated Meritt's own ideas about how to "approach the AAZPA's national breeding plan," and he recognized that zoos had much to learn from the wildlife biologists that would be relevant to captive propagation. Equally important, the opening of lines of communication between the FWS and the AAZPA promised to turn the master breeding plan into an initiative that would receive input and support from the FWS. Meritt concluded that "we are at a definite crossroads in our relationship with USDI-FWS and [I] am more convinced than ever that we are in an extremely good position." Meritt turned out to be correct in this assessment, and by the mid-1980s the AAZPA and FWS had initiated several cooperative captive-breeding programs. This cooperation pointed to the value of the "long and hard" work that AAZPA members such as Meritt had done to quietly and rationally persuade the FWS that the AAZPA was a scientifically oriented conservation organization, not just another interest group jockeying to influence FWS policy. As Meritt said, "We are professionals and we are being recognized, finally, as such."[23]

Breeding endangered animals was not an entirely altruistic activity, because it justified zoos' continued acquisition of those species, but the AAZPA had historically supported a range of conservation legislation, and it continued to do so. As essential as breeding programs were to zoos' survival, a deeper commitment to conservation demanded that zoos defend the ESA's basic principles against the growing number of critics in the late 1970s. Attacks on the ESA's integrity actually gave the zoo community the opportunity to work cooperatively with some of the conservation and animal welfare organizations that had been occasional adversaries. Thus, even while working to establish its SSPs, the AAZPA began to argue before Congress that much more needed to be done to protect endangered animals.

Blakely had suggested reaching out to moderate conservation groups, and in 1976 Wagner and ZooAct leaders organized the first Zoological/Environmental Conference, with representatives from the Audubon Society, the National Wildlife Federation, the Sierra Club, and humane groups. This dialogue continued over the next few years through regularly scheduled meetings that added representatives from Friends of the Earth and the Environmental Defense Fund, and featured such topics as CITES, ESA reauthorizations, the pet trade, and captive breeding programs. Meeting first on neutral ground at the Dupont Plaza Hotel, in Washington, D.C., the participants opened a "positive dialogue" about some of their shared perspectives on conservation issues. Although Steele and Wagner could not convince the group about the validity of the AAZPA's position on injurious wildlife regulations and killer whale permits, they did make headway in conveying their commitment to the ESA. A few months later, Wagner initiated a second conference, this time held at the National Wildlife Federation headquarters, and in April 1978 the participants met a third time at the HSUS's Washington office. The purpose of this final meeting was to rally support for section 7 of the ESA, a provision requiring federal agencies to ensure that their actions did not jeopardize endangered species. Wagner felt that by expressing its opposition to amendments weakening this provision, the AAZPA could "strengthen our relationship with the environmentalists."[24]

By the late 1970s, the ESA had become a highly controversial piece of environmental legislation.[25] By requiring the conservation of endangered species, the ESA essentially provided for habitat protection, a provision that had probably seemed fine to politicians in 1973, who imagined that tracts of Alaska or other such wilderness areas might be preserved for the benefit of charismatic animals like grizzly bears. But the list of endangered animals grew each year, and before long (1977), the Supreme Court had ruled that Louisiana could not construct a highway through critical sandhill crane habitat. Then, biologists discovered that a dam in Tennessee threatened the snail darter, a three-inch fish that became the poster animal for what was "wrong" with the ESA. Surely, politicians argued, they had not intended for the ESA to hold development hostage to the protection of such insignificant creatures. At the 1979 reauthorization hearings, Senator Howard Baker (R-TN) stated, "We who voted for the ESA with the honest intention of protecting such glories of nature as the wolf, the eagle, and other treasures have found that extremists with wholly different motives are using this noble act for meanly obstructive ends."[26] The political significance of the senator's remarks were made clear by a contemporary public opinion survey that found low levels of knowledge about the snail darter controversy (17 percent) and widespread opposition to preventing "essential" development to protect such species. Instead, overwhelming majorities favored the protections of attractive animals such as bald eagles and mountain lions that had some historical or economic importance.[27]

Each year the ESA has been up for reauthorization, its opponents have tried to add weakening amendments, and, beginning in 1978, the AAZPA initiated a coalition of environmentalists in actively working against such amendments. In that year's Zoological/Environmental Conference, Wagner discussed AAZPA plans to mobilize its members and zoo visitors against a 1978 amendment that would have stripped the ESA of its power to designate "critical habitats" necessary to preserve highly endangered species. AAZPA members wrote letters to their congressional representatives, and Wagner testified before a House subcommittee to oppose the proposed amendments, actions that drew "acclaim" from the previously somewhat skeptical environmental and conservation community.[28] The ESA's defenders won this first reauthorization battle, but the ESA's opponents shortened the authorization period from five to two years, which meant that conservationists and environmentalists would need to devote nearly full-time energy to assuring the future of this vital legislation.

Following the 1978 reauthorization, the AAZPA geared up almost immediately for the next fight, joining with its new allies to develop positions on proposed weakening amendments. ZooAct offered its assistance by organizing an "urgent mailgram campaign" to solicit letters of support for President Carter's stated intention to veto a Tellico Dam amendment designed to circumvent the ESA. At the local level, a few zoo directors followed the AAZPA's lead and reached out to "legitimate conservation organizations" to publicize zoos' role in conservation and education, or to form more significant liaisons to preserve the ESA.[29]

The unity of this coalition was challenged in 1979, however, when Chairman Breaux, of the House Subcommittee on Fisheries and Wildlife Conservation and the Environment, introduced several amendments that would have placed the independent Endangered Species Scientific Authority under the Secretary of the Interior. Some environmentalists predicted that this move would effectively weaken the ESA by subjecting decisions about international trade subject to political pressures, but the AAZPA wanted to avoid a fight with Breaux, who had been friendly to the zoo community in the past and whose committee had jurisdiction over "literally every wildlife/conservation bill." As an interest group that came before Breaux's committee frequently, the AAZPA desired to maintain a positive relationship with this powerful chairman. The AAZPA persuaded the other coalition members that neither it nor they could afford to alienate Breaux, because they would undoubtedly require his support in the future. In the end the AAZPA felt that the coalition deserved some credit for the House's speedy consideration of the reauthorization.[30] Perhaps more important, the AAZPA had demonstrated to the conservation community that it was an effective ally in the struggle to defend environmental legislation, a role that it would continue to play in the decades ahead. Even though the AAZPA needed to protect members' right to import and exchange endangered animals, its sincere commitment to conservation led it to establish a closer working relationship with other conservation organizations and assume an active role in lobbying Congress during ESA reauthorizations. In addition to helping zoos fulfill their stated conservation mission, these collaborations created a positive image for zoos and cemented their relationships with conservation organizations and the FWS, both of which had been troublesome adversaries in the early 1970s. The AAZPA leadership recognized that, if its member zoos hoped to avoid negative publicity, they would have to support the kinds of policies and standards that might put such places as Great Adventure out of business.

Bringing Them In

Just as the AAZPA found common ground with conservation organizations regarding the ESA, so too it worked alongside animal welfare organizations to strengthen regulations governing the transit of animals. During congressional debates about the Endangered Species Conservation Act of 1969, the AAZPA had complained about inadequate shipping standards for the transportation of wild animals, and the issue continued to attract intermittent attention. Well into the 1970s, airlines lost many animals in transit both because they allowed shippers to use flimsy, poorly ventilated containers packed full of animals and because they (and shippers) failed to provide adequate care. Crated animals were sometimes left for long stretches without food or water in hot and poorly ventilated areas.[31] The deaths of animals at the hands of shippers and airlines were often grisly. In one example, a Miami dealer shipped two sloths to a Nebraska zoo in a smooth-sided box without anything for the animals to grip. The sloths arrived dead, killed by "ghastly wounds inflicted on each other in their attempts to find a solid handhold." The deputy director of airport activities for the Washington Humane Society described going to the National Airport in June 1973 to take possession of more than eight hundred rats, most of which were dead, dying, or being eaten by the few survivors. Another dealer shipped a crate of animals to the Detroit Metropolitan Airport; when the shipment arrived, airline workers smelled dead animals and called Dorothy Dyce, the AWI's laboratory animal consultant. Dyce took the crate to the Detroit Zoo, where she and the zoo veterinarian discovered, among other animals, "12 horned frogs, 9 dead; 20 rainbow lizards, 18 dead; 25 frogs, 20 dead; one raccoon dead; 25 baby alligators, five dead."[32] Such incidents had outraged animal welfare groups since the late 1960s and led them to lobby Congress for tighter shipping regulations.

Animal shippers and commercial carriers essentially operated without any government oversight of the conditions of transport. The AWA gave the USDA power to regulate the care of animals only in permanent facilities, and no other federal agency had rules governing the welfare of animals in transit. In December 1973, a subcommittee of the House Committee on Government Operations held hearings on this issue and recommended creating an interagency committee of the USDA, Civil Aeronautics Board (CAB), and the Federal Aviation Authority to work out a jurisdictional solution.[33] By mid-1974, Congress took steps to address the problems of shoddy shipping crates, lack of temperature control in airplane cargo holds, and inadequate care for animals at airline terminals. Representative Tomas Foley (D-WA), chair of the Subcommittee of Livestock and Grains of the House Agriculture Committee, introduced the Animal Welfare Amendments of 1974 (HR 5808) to force shippers, airlines, and veterinarians to assure the safe air transport of animals. In 1975, Senator Lowell Weicker (R-CT) introduced S 1941—aside from the weaker language regarding veterinarian inspections, this amendment resembled Foley's in bringing "airline and terminal facilities under regulations of the Animal Welfare Act" and directing the "Secretary of Agriculture to promulgate strict standards for the care of animals in transit, including standards for containers, feed, water, ventilation, temperature, and veterinary care."[34]

Zoos, whose animal imports constituted only a tiny percentage of the animal trade, could not afford the kinds of losses incurred because of weak shipping regulations, so the AAZPA cautiously supported congressional efforts to protect animals during transit. The issue captured legislators' attention partly because pet owners around the country made emotional pleas to protect their loved animals, and partly because some of the animals were destined for publicly funded institutions. In testimony before a 1975 Senate hearing, one witness noted that "zoo animals are also usually paid for with taxpayer's money," as were animals bound for research projects funded by federal grants. She reasoned that, because "taxpayers [were] already subsidizing some of the airlines to the tune of 68 million dollars in 1974," they should receive better treatment for the animals they shipped. Zoo leaders agreed, but their work on behalf of creating better standards was tempered by the recognition that in the face of "complex regulations," the airline transport industry might refuse live animal shipments altogether.[35]

The USDA and its Animal and Plant Health Inspection Service (APHIS) and affected industries fought a losing battle against S 1941. The USDA preferred that all interested parties, which included zoos, come up with "standards and procedures" that could be adopted on a "voluntary basis." Phil Campbell, Undersecretary of Agriculture, pointed out that CAB was currently holding administrative proceedings on the very issue of animal welfare, so legislators should wait to see what guidelines resulted. APHIS, despite its role as the official protector of animal welfare, also failed to support the legislation. Dr. Pierre A. Chaloux, assistant deputy administrator of veterinary services at APHIS, communicated his agency's philosophical resistance to legislation that "substantially increases the regulatory power of the Federal Government." Instead, Chaloux seconded the USDA's recommendation that the parties adopt voluntary standards and general guidelines issued by CAB. This insistence that CAB could solve the problem was undercut by Katherine Kent, assistant director of CAB's bureau of economics, who testified that CAB lacked both the plenary power and the necessary expertise to develop such regulations for the humane air transport of animals. Though assuring Congress that CAB's investigations of the problem would continue, Kent warned that animals would be protected in transit only when a single agency was given clear jurisdiction over both the shippers and the airlines. This agency, she argued, should be the USDA, because it alone employed the veterinarians and field officers with the expertise to develop and enforce humane transport rules.[36]

In response to this stonewalling, Senator Weicker rhetorically asked, "Why all the opposition in the bureaucracy?" The answer, it appeared to him, was that the bureaucracy was looking out for the interests of the airlines, the pet industry, the shippers, and the biomedical research companies, all of which objected to the "economic impact" of humane shipping rules that would, for example, increase the amount of space required for animals.[37] Leo Seybold, the Air Transport Association (ATA) representative at the hearing, explained that the airlines had "taken the initiative" to "improve conditions" for animals traveling by airline through their tariff procedures, but shippers had protested the cost of the suggested improvements. Of course, the ATA had its own cost concerns and opposed any actual regulation of animals in transit because the USDA did "not have the knowledge to determine what is or is not economically practical or technologically feasible with regard to aircraft." The ATA also tried to shift the responsibility to shippers, arguing that the airlines were already regulated too much.[38]

The medical research industry representatives echoed these economic arguments. Frederick N. Andrews, vice president for research in the graduate school of Purdue University, objected to the clause specifying that only animals determined "sound and healthy" be delivered; he claimed that the burden of inspecting individual animals in shipments often numbering in the hundreds would be "of questionable value in relation to the increased costs of such animals to research facilities."[39] Similarly, the Academy of Pharmaceutical Science of the American Pharmaceutical Association, an interest group with both academic and industry members, wanted a drastically watered-down bill with major exemptions for research animals. In a prepared statement stressing the economic impact of the legislation, the academy informed Weicker's committee that the increased steps in veterinary care, transportation, handling, and record keeping would lead to higher costs for consumers. Moreover, the academy claimed that industry, not the USDA, could best write the standards for the humane "care, handling, treatment, and transportation" of animals.[40] As this testimony made obvious, rules written or influenced by the pet and medical research industries would never adequately protect zoo animals, because the industries shipped animals in bulk and could afford substantial losses.

As the legislation worked its way through Congress, it appeared as though the threat of regulation might push the airline industry to stop carrying any zoo animals at all. Facing the potential end to one of the primary conduits for animals for zoos and uncertain about S 1941's fate, the AAZPA took the advice of the Department of Agriculture and approached the airlines about creating voluntary guidelines for shipping animals safely. They also put together a manual that detailed how to ship various kinds of animals. The necessity of such outreach was evidenced by the airlines' reluctance to deal with zoo animals. Air cargo clerks and baggage handlers resented the extra effort they made to keep such animals healthy. One can sympathize with the baggage handlers to a certain extent because sometimes the animals "lost" in the cargo space were intimidating. A snake bound for the North Carolina Zoo, for example, slithered out of its cage, forcing airline employees to find and catch it. In response to this additional workload, employees sometimes avoided boarding zoo animals by falsely claiming that their airline refused to import certain animals or had policies that limited the number of animals they imported. American Airlines employees, for example, refused to ship a black lemur and a pygmy marmoset to a zoo, even though they had no policy of refusing to ship nonhuman primates.[41] In other cases, airlines legally secured the right to refuse animals in their cargo by writing this provision into their tariff rules, which were part of the contract between the airline and the shipper.

Combining cooperative and coercive strategies, and hedging their bets against the possibility that congressional action would fail, the AAZPA simultaneously took their case to the CAB, testifying before an administrative law judge about the airlines' continuing "embargo" or "restriction of certain kinds of animals," particularly "snakes and large animals." The AAZPA and ZooAct hoped to convince the judge that the carriers would "retreat from their position that properly packaged specimens cannot be transported by air." From the AAZPA's perspective, the passage of S 1941 would make it illegal for airlines to refuse their cargo.[42] Both amendments passed their respective chambers, but in the end Foley's amendment was tabled in favor of Weicker's, which became law on April 22, 1976. Ultimately successful in both their appeal and the AWA amendment, the AAZPA continued working with airlines on guidelines for shipping zoo animals.

The AAZPA's involvement in the transportation issue not only was a natural extension of its efforts in the 1960s to save exotic animals languishing in USDA quarantine facilities, but also marked a convergence of zoos' interests with those of animal welfare organizations. The latter focused largely on the pets and laboratory animals, which constituted the bulk of the commercial carrier trade, and their vocal presence during congressional debates on the issue no doubt lent power to the AAZPA's smaller-scale lobbying efforts. Congressman Whitehurst reported that numerous letters from animal lovers around the country convinced him that the government needed to act on the problem, and the legislation itself emphasized protections for pet animals.[43] The AAZPA's work on this issue demonstrated to the animal welfare community that, even though zoos sought regulations to benefit their members, they simultaneously protected animals beyond their gates. Nonetheless, if the ESA implementation and air transport experiences revealed that zoos and animal protectionists could work together on some shared goals, issues involving the capture and display of marine mammals revealed that the AAZPA could not count on support from animal protectionists on marine mammal issues.

Marine Mammal Controversies

The implementation of the AWA amendments, the MMPA, and the ESA significantly influenced zoos, much as Zeehandelaar had predicted in 1973, when he warned that "normally the danger" rests not in the laws themselves but in the "proposed implementing regulations." Relatively new to federal bureaucracies, it took zoo professionals several years to learn how rules were made and enforced and how they could be changed. Zoo leaders initially also feared that animal welfare and environmental groups would turn the rule-making process to their own advantage. Eventually, however, they learned how to use the bureaucracy to their advantage. Even by 1977, Wagner could boast that "we have established a much closer working relationship with the governmental agencies which have regulatory responsibilities" over zoos and aquariums. He underlined the importance of this relationship by reminding members that zoos still had "much catching up" to do "if we are to assume the role of *the* authority on the management of captive wildlife."[44] Wagner knew that, even with increased federal regulation, zoos could emerge more powerful than before if they played an active role in writing the rules that set standards for the capture and care of wild animals.

Because APHIS and other regulatory agencies often lacked the technical expertise to evaluate the health or welfare of exotic animals, zoos shaped a significant number of the rules under which they now operate. Essentially, zoos' economic interest in keeping animals alive and healthy made them experts in the capture, transportation, feeding, housing, and veterinary treatment of exotic species. To stay in business, zoos had to ensure the very animal welfare that, at times, their critics accused them of neglecting. It is not surprising that we find that government agencies often turned to the zoo community when developing rules governing the capture or care of wild animals. Initially, however, the AAZPA's relationship with these agencies was rocky and its members' participation in the rule-making process was inconsistent.

The multiyear process of developing rules regarding the care and capture of marine mammals nicely illustrates the difficulties zoos faced when learning how to negotiate the federal bureaucracies. Many zoos not only had seals, sea lions, porpoises, and other marine mammals on display, but also were eager to expand these displays to take advantage of the public enthusiasm for such animals. The AWA and MMPA, however, had complicated the acquisition of marine mammals by dispersing power throughout multiple regulatory agencies. Under the AWA, APHIS had the responsibility to develop standards of care for marine mammals. A special Marine Mammal Commission, however, also had input on these standards. Most important, the issue of how to implement permit-granting aspects of the MMPA (1972) was being worked out by the Marine Mammal Commission and a Committee of Scientific Advisors on Marine Mammals, in consultation with the Department of Commerce, the Department of the Interior, the National Marine Fisheries Service, the National Park Service, the Naval Undersea Center, the U.S. Navy, the National Science Foundation, and other affected government agencies.

Such a list might have scared off all but the most seasoned lobbyists, and sure enough, when these actors met at a three-day conference in early 1975, the only zoo to send a representative was the San Diego Zoo.[45]

Although it took some time for many member zoos to figure out how to participate, the AAZPA and ZooAct leadership were active in the marine mammal issue from the beginning. In October 1975 several AAZPA and ZooAct representatives testified at the MMPA legislative oversight hearing held before Leggett's House Merchant Marine and Fisheries Committee Subcommittee on Fisheries and Wildlife Conservation and the Environment. Zoo community witnesses Wagner, John Prescott (the New England Aquarium), Steele, and Lanny Cornell (Sea World) urged amendments to the MMPA that would have made it easier for zoos to take animals from the wild. Steele and Cornell, representing the ZooAct wing of the zoo community, portrayed the permitting process as having been hijacked by "a handful of humaniac extremist organizations." The "hysterical dictating by a vocal and emotional few" hampered zoos' ability to collect and display marine mammals as Congress intended, but also prevented the acquisition of scientific knowledge about animals, thereby threatening to return earth "to the dark ages." Because the regulatory agencies were intimidated by these groups, they dragged out the application process and issued only "unreasonable restrictive" collection permits. Steele quoted a Sea World employee who complained about a collection permit allowing the aquarium to "capture any killer whale as long it is not black and white." To weaken the protectionists' obstructive power, Steele recommended amendments allowing "approved display institutions" to obtain "open-ended permits" to capture certain species as needed. Short of an amendment, Steele urged more training of federal personnel regarding husbandry and collecting procedures, and he graciously offered the zoo community's assistance in that regard.[46] Wagner, Steele, and Cornell essentially felt that the extended public comment on permit applications empowered the protection groups and subverted Congress's intention to allow zoos and aquariums to capture marine mammals for display purposes.

Political opposition to the taking of marine mammals, particularly killer whales, provided ZooAct with an opportunity to demonstrate its value to the zoo community. Animal protection groups had never been happy with MMPA language that allowed capture permits for public display purposes. In 1976, Sea World's attempted capture of several killer whales in Puget Sound outraged many and led to an out-of-court settlement in which Sea World agreed not to collect any killer whales in Puget Sound, though its permit allowed it to do so. Unsatisfied with this short-term solution, two Washington state Democratic members of Congress, Representative Don Bonker and

Senator Warren Magnuson, introduced legislation (HR 12460 and S 3130) amending the MMPA to place a moratorium on the capture of killer whales for display purposes.[47]

The zoo and aquarium community's participation in House hearings on the killer whale amendments revealed the underlying vulnerability of the AAZPA's commercial members, the growing political strength of the AAZPA and ZooAct, and the significance of the AAZPA's emphasis on public education and scientific research. Debate at the hearings centered on the commercial status of Sea World and other aquariums. Bonker insisted that his amendment protected killer whales from capture for "primarily commercial purpose." Project Monitor's president, Milton Kaufmann, whose group kept tabs on marine mammal legislation portrayed the amendment's opponents as "the spokesmen for corporations who see their profits threatened." Friends of the Earth's conservation director, Tom Garrett, similarly characterized aquariums as an "industry" that had to "pay a claque of lobbyists to wet the floors of congressional hearing rooms with expensive tears." Toby Cooper, wildlife programs coordinator for Defenders of Wildlife, argued that only a moratorium could prevent aquariums from exploiting the "commercial successes of killer whales." And Christine Stevens, from the Society for Animal Protective Legislature, decried Sea World's "extreme commerciality."[48]

The protectionists' attempt to portray aquariums as commercial interests was received skeptically by Chairman Leggett, Congressman Oberstar, and a host of congressmen who testified against the amendment. The committee members frequently returned to a theme from the initial MMPA debates: the crucial public education role of aquariums. By 1976, the political utility of this theme was obvious to the AAZPA, and Wagner put it at the center of his statement: "We are before you today to once again define our role as educators of the public regarding wildlife conservation." Each zoo witness provided statistics attesting to the number of citizens reached by zoos and aquariums, while Leggett and Oberstar used friendly follow-up questions and comments to highlight zoos' educational role. Oberstar, for example, asked Wagner whether "it [was] a fair statement that visitors to zoos have, in large part, accounted for public pressure to protect certain endangered species?" Representatives from Sea World's congressional districts expressed opposition to the amendment based on their firm belief in aquariums' educational value. Lionel Van Deerlin (D-CA) argued that killer whales were "in greater danger *before* Sea World began exhibiting them for an admiring public," and Clair Burgener (R-CA) agreed that Sea World played a major role in getting information about killer whales "to the public." Their Republican colleague from the San Diego area, Bob Wilson, dramatized the personal encounters with wildlife possible at aquariums. He described "eyeballing" and "patting" a captive killer whale who "smiled back at me," and he hoped that millions of Americans would continue to be able to see such creatures on display. As in the 1971 MMPA hearings, concerns about the commercial nature of aquariums were effectively drowned out by repeated claims about the educational mission of those aquariums.[49]

Although the commercial aquarium representatives were immediately threatened by the amendment, AAZPA representatives Wagner, Charles Hoessle, and Prescott emphasized that a prohibition against taking killer whales would set a "precedent" for additional prohibitions that could affect many more zoos. Hoessle, the St. Louis Zoo's deputy director, testified that, though his zoo had no desire to exhibit killer whales, it "strongly opposed setting a precedent," fearing that "every additional regulation affecting the flow of live animals to zoos and aquariums detracts from the scientific and educational function of the institution." The committee members expressed sympathy for this line of argument, and in cross-examination encouraged Wagner and Prescott to reiterate and elaborate upon their fear. Leggett, who clearly opposed the amendment, pushed the amendment's supporters to explain why killer whales, of all marine mammals, should receive this unique protection. His underlying point, which coincided with the zoo community's perspective, was that, in the absence of ESA status, such a protection would be unscientific and would open the door for a wholesale extension of restrictions never intended in the original MMPA.[50]

The opposition to the amendment was significant. The zoo community enjoyed supportive testimony by congressmen from aquarium districts in Florida, California, and Illinois. The affected government agencies—National Oceanic and Atmospheric Administration the Department of Commerce, and the Marine Mammal Commission—and their witnesses opposed the amendment. A series of government researchers concluded that no scientific evidence supported the need to protect killer whales in this way, and statements by Wagner and other AAZPA members affirmed aquariums' conservation and education role. Leggett was clearly unsympathetic to the bill, and it was never reported out of his subcommittee.

ZooAct played a major role in the successful dissipation of the killer whale amendment threat. As expected, Steele provided a written statement from ZooAct and appeared in person at the House hearing. ZooAct's more substantial contribution came behind the scenes. First, Sea World's economic significance to the San Diego region made it easy to mobilize support at the hearing from several local congressmen. Second, in the weeks prior to the hearing, ZooAct's president, Frank Powell, initiated a grassroots campaign at three Sea World parks to collect signatures and letters in opposition to the killer whale amendment. More than 10,000 letters and 90,000 signatures were delivered in a photo-op ceremony to Robert Leggett and three friendly congressmen—Bob Wilson, Lionel Van Deerlin, and Clair Burgener. Leggett applauded this "impressive" campaign and opined that "it was about time that we heard from the other side" (i.e., the zoo side).[51] ZooAct had demonstrated that the financial muscle of the AAZPA's commercial members could, as Braker had predicted several years earlier, greatly assist the poorer public zoos by vigorously defending the entire zoo community from additional regulations.

During this time, Wagner had to give equal attention to MMPA developments on the administrative side. In mid-1975 he proposed the creation of a subcommittee to attend all four of the Marine Mammal Commission's annual meetings and represent the AAZPA position on marine mammal husbandry issues.[52] This activity was crucial, because even aquariums that already held marine mammals would be affected by new care and handling standards. When APHIS published proposed standards and regulations for "the humane handling, care, and transportation of marine mammals" in 1977, the AAZPA participated in several working sessions with USDA officials and tried, unsuccessfully, to write in a grandfather clause for existing facilities, though it did manage to insert draft language giving noncompliant facilities (estimated to be 75 percent) a "reasonable" amount of time to make required changes.[53]

The rule-making process dragged out over four years, and during this time the AAZPA discovered that it could use congressional allies to force slow-moving government agencies to respond to zoos' needs. With the help of Congressman Leggett, the AAZPA dragged a reluctant USDA to a two-day closed conference in 1978 to discuss progress on the development of reasonable rules.[54] It is interesting to note that the AAZPA sought to protect its members not just with favorable standards, but also by preventing traveling acts (e.g., circuses) from receiving the same protections as established zoos and aquariums. The AAZPA clearly preferred standards that only its members would be able to meet easily. It was in zoos' interest to make life difficult for "traveling acts" because those acts attracted so much of the animal welfare groups' ire. If government regulations could put them out of business, zoos would be able to claim a moral high ground.

In 1979, after what Steele described as a "four-year struggle," the USDA released its final regulations on the collection, transportation, and maintenance of marine mammals. Although these final rules did not incorporate every AAZPA recommendation, Steele believed that the AAZPA's active participation in the rule-making process put in "in the enviable position to incorporate needed changes" in future rules. Such rule-making cooperation occurred almost immediately. Before the end of 1979, the USDA asked several AAZPA and Sea World representatives to "participate in a training session" of marine mammal facilities inspectors. As part of their training, inspectors would visit Sea World's Orlando, Florida, facility. In his subsequent report to members, Wagner noted that further and "exacting" revisions of the marine mammal care provisions were likely and "will obviously require adjustments on the part of AAZPA members, *as well as inspectors for Agriculture*" (emphasis added). After nearly a decade of active engagement with these governmental agencies, the prospect of further revision of marine mammal care standards was faced calmly and optimistically rather than with concern and pessimism.[55]

The AAZPA experience with the marine mammal issue resembled its struggle with the ESA implementation in its broad outline. In both situations, Wagner ensured that the AAZPA presented the zoo community's concerns to both legislative and administrative bodies. Steele, in his dual capacity as ZooAct's director and the AAZPA's legislative representative, worked closely with Wagner to articulate the zoo community's interests through a variety of techniques, including grassroots lobbying. Although different government agencies were at the center of each issue, Leggett's subcommittee held all hearings dealing with the ESA and the MMPA, and he proved a useful congressional ally for zoos. At a deeper level, congressmen were receptive to the argument that the ESA and MMPA were intended to allow and even encourage the display of animals. By stressing the conservation, educational, and research missions of zoos, Wagner appealed directly to language in both acts. By simultaneously working to professionalize the industry, Wagner assured that an increasing number of zoos were fulfilling that three-part mission. Funding that mission, however, remained a challenge.

Federal Funding and Zoos

Even as the AAZPA enjoyed some successes in the regulatory and legislative arenas, many of its member zoos, dependent upon local tax bases, were struggling to fund the improvements required by new federal laws. The decade leading up to 1975 had been one of turmoil and change not just for zoos, but also for American cities in general. The civil rights and women's rights movement, among other factors, profoundly affected the political landscape. The federal government had initiated numerous social welfare programs designed to alleviate urban poverty; white residents fled cities for the suburbs, as did many industries, leaving behind a diminished tax base; and global competition, particularly in the auto industry, led to massive job losses in urban areas. In short, American cities were reeling. For public zoos, particularly those located in midwestern and northeastern cities, these changes reduced tax support and gate revenues at a moment of critical need. By the mid-1970s, zoos required far more than continued operating support;

they needed to expand in dramatic and costly ways. External pressure from the AWA, the ESA, and the AAZPA created the expectation, indeed the requirement, that public zoos reinvent themselves: older, sometimes inhumane cages needed to be replaced with modern, professionally designed exhibit spaces that improved the animals' welfare and enhanced visitor learning; full-time veterinarians were needed to care for animals; education and research missions had to be expanded; and the management structure had to be professionalized to meet AAZPA accreditation requirements.

Recognizing its members' increased financial burdens, the AAZPA looked for funding solutions from the federal government. It used the newsletter to identify a wide range of federal grants for which zoos could apply and to publicize its members' successful grant applications. When the federal government created a new grant program for museums, AAZPA leaders gained zoo representation on its advisory board. Finally, as education grants became available, the AAZPA coached members and the government alike on zoos' educational mission.

The AWA amendments of 1970 threatened to close zoos that did not make fundamental changes to meet minimum standards. In public the AAZPA tended to portray roadside zoos as the main offenders, but public zoos also had difficulty meeting the standards required by the AWA for an operating license. APHIS refused to issue such a license for one AAZPA member, the Beardsley Park Zoo (Bridgeport, Connecticut), until the zoo built a new animal hospital, retrained its staff, and made many other changes to the physical facility. In 1975, APHIS reported that a hundred roadside zoos had closed because they could not afford to meet new standards of animal care, and one hundred twenty larger zoos had invested in improvements of various kinds, including major ones such as new enclosures and professional staff. The money for such improvements, however, was not always forthcoming from local government. When the Roger Williams Zoo (Providence, Rhode Island) had to undertake some drainage and fencing construction to qualify for its USDA exhibitor's license, the mayor of Providence refused to "sign the required license without a written guarantee that the Government will not require that the city spend any more money at the zoo." In this case, local government not only failed to support, in principle, the spirit of the AWA, but also could not afford to support it, partly because of a depressed tax base and partly because it was constrained by municipal law making it illegal for the zoo to charge an entrance fee.[56]

Zoo leaders and their political allies stressed this connection between new government regulations and zoos' economic woes as a central reason to support Senator Mark Hatfield's 1974 Zoo Assistance Act, which promised millions of dollars in grants and loans. Witness after witness at the subcommittee hearing on this bill bemoaned the sorry state of zoos finances. The first speaker, Congressman Whitehurst, who had introduced a companion bill in the House, established this theme when he described zoos as "struggling for survival due to a lack of funds." It is telling that Senator Hatfield admitted that the federal grants would not likely stimulate matching money from local government because the basic problem was that local government lacked money to lift zoos from their "antiquated and decrepit" condition. Prescott was even more blunt: "Cities for years have considered zoos and aquaria as recreational frills. Their budgets for these facilities have remained static, our personnel have been grossly underpaid, and capital improvement programs have remained under funded." Sophie Danforth, president of the Rhode Island Zoological Society, echoed this sentiment when she complained that "zoos begin with a Z, the end of the alphabet. And that's where they are when it comes to a city budget." In Rhode Island, this general funding problem had been made worse by the loss of a naval base and the flight of the textile industry. Prescott reflected the generally pessimistic mood by concluding that the "future forecast is gloomy. Cities face larger and larger deficits, and individual private donations are becoming smaller and harder to obtain."[57]

Zoo people were quick to assure legislators that they sought federal support only because animals suffered when zoos were underfunded. Such a concern reflected the bill's underlying purpose: to eliminate inhumane conditions in zoos through a set of financial incentives. In fact, Whitehurst acknowledged that "increased governmental regulation of the conditions in zoos" had forced the adoption of "new and often costly concepts in zoo designs" that local governments could ill afford. The director of the Milwaukee County Zoo noted that, although zoo employees often had "good expertise concerning the daily care, sanitation, health, and medical attention" required by their animals, they often lacked "the funds necessary to put their expertise fully to work." The director of another midwestern zoo believed that federal funding "may well prove to be one of the only ways zoos may make changes to meet minimum standards promulgated by recent legislation."[58]

Similarly, the expansion of the educational and research missions of zoos in response to the ESA and the AAZPA caused financial stress and justified increased support. Again, Whitehurst established the theme by arguing that zoos no longer simply entertained visitors but performed "an important educational function" and "important services in conservation and research." Hatfield elaborated by explaining that zoos were developing sophisticated breeding programs that required more naturalistic animal enclosures than the "concrete slabs" common in many facilities. Zoo spokespeople eagerly embraced these three rhetorically familiar roles—education, research, and conservation—but they stressed that each would require a significant public investment. William Conway, director of the New York Zoological Society, explained that the zoo "constitutes a unique medium for the dissemination of environmental education." Prescott added that "today's educators increasingly utilize our facilities for off-campus learning, and the academic community views our valuable collections as reservoirs of knowledge generally unavailable to the laboratory sciences." Danforth noted that, "given the proper facilities and well-trained personnel, any good zoo can play its part in saving an endangered species." Zoos, however, needed financial assistance if they were to become something more than "recreational frills." They had the desire and potential to become research institutions on par with universities, but their ambitions were restricted by "a lack of available funding." Quite simply, they needed government support "equal to that given other scientific institutions."[59]

As we have seen, the federal assistance program for zoos never got out of subcommittee, but zoos probably profited from their enforced "independence" because it led them to broaden their funding base in ways that proved useful later. Even while zoos argued for federal assistance in 1974, a few were already tapping into other science-oriented grants programs. Director Conway acknowledged that the Philadelphia and New York zoos, which had well-established research programs, had "a long history of funding" from the National Science Foundation and the National Institutes of Health. And, although Whitehurst claimed that zoos "have had an increasingly difficult time competing for funds with higher priority social programs," in fact zoos soon began applying for *social program* funds.[60] By the mid-1970s, zoos discovered that they could successfully apply for grants from a wide variety of federal programs designed to solve urban poverty and redevelopment problems that had little to do with the welfare of captive animals.

One case in particular serves as an excellent example of how zoos broadened their funding base by taking advantage of multiple federal programs. In 1976, the Buffalo Zoological Gardens hired a full-time grants/development director to help the institution meet its "severe need for capital improvements" as well as its expanded "conservation, science, and education" mission. The investment in a grant director paid significant dividends. Within two years, the zoo had received $155,000 from the City of Buffalo's community development block grant (a federal assistance program run by the U.S. Department of Housing and Urban Development); $238,150 from the Bureau of Outdoor Recreation (under the Department of the Interior); $309,000 from the Economic Development Administration's Local Public Works Capital Development and Investment Program; $234,244 from the Comprehensive Employment and Training Act (CETA, part of the U.S. Department of Labor); and $375,000 in appropriations from New York state.[61] Nearly all of these programs targeted urban poverty and decline in one way or another, either by providing jobs or funding rebuilding. Not one was designed specifically to meet the needs of animals, but a good grant writer, with support from local politicians, could easily make the case that any one of those program's larger purpose could be fulfilled within the context of the urban zoo.

Other zoos pursued similar antipoverty funding opportunities during this period. Staffing shortages, a frequent complaint, could be addressed through CETA—a social program set up under the Nixon administration in 1973 to help alleviate poverty by providing grant money for job and educational programs.[62] Local public institutions received funding to hire additional workers from underemployed populations, particularly youths. For example, in 1975 the Roger Williams Park Zoo—which had bemoaned its paltry budget at the Senate hearing on the zoo assistance board—nearly doubled its zoo staff when it received CETA money for ten new staff members, including an educational coordinator and an exhibits and graphic arts director. Similarly, the Overton Park Zoo and Aquarium (now the Memphis Zoo, in Memphis, Tennessee) added thirty employees recruited from economically deprived areas and funded by a CETA grant. Such federally funded employees were not always the "highly trained professionals" that zoos begged for at the Senate hearing. For example, the Como Zoo (St. Paul, Minnesota) had cobbled together ten federal, state, and local grants by 1975, but these grants funded only untrained employees, mainly youths and high school dropouts. Moreover, the capture of federal dollars meant the loss of city dollars, because the Como Zoo saw its city funding cut in proportion to its federal gains.[63] Thus, antipoverty programs could have mixed effects: they allowed zoo staffs to grow, but often the growth was in unskilled laborers, not in much-needed professionals. Grants broadened a zoo's funding base, but did not necessarily deepen it. Nonetheless, the process of applying for and administering such grants forced zoos to articulate their public mission. As Wagner reminded AAZPA directors in 1978, competing for federal grants was "an excellent opportunity for others to find out that zoos and aquariums and the AAZPA are no longer sleeping."[64]

In addition to getting funding for human resources, zoos also took advantage of federal programs for capital improvements. Grants from the Public Works Employment Act of 1976 made money available through the Secretary of Commerce to state and local governments for public works projects including "new construction, renovation, or repair" of "local public works projects." The bill's stated aim was to "provide employment opportunities in high areas of unemployment through the expeditious construction or renovation of useful public facilities and provide a stimulus to the national economy." The Franklin Park Zoo, in Boston, assembled $4,895,000 from this fund, a state bond authorization, other grants, and private donations. The money funded an "African Tropical Forest Pavilion" among other projects in a master plan developed by the Boston Zoological Society. Similarly, between 1975 and 1978 the Dickerson Park Zoo, in Springfield, Missouri, spent $245,000 in federal community development funds to restore eroded shoreline and make other improvements. A particularly lucrative grant program came from the Department of the Interior's Bureau of Outdoor Recreation (later the Heritage Conservation and Recreation Service). The Oklahoma City Zoo was among the major beneficiaries, receiving $291,000 in 1978 to expand its facilities by fifty acres (more than 10 percent) to house a new exhibit for "large African animals." Wagner kept the membership informed about such funding opportunities, and he lobbied the various granting agencies to include zoos among their eligible applicants.[65]

In all of these cases, federal money allowed zoos to expand and modernize their exhibits, thereby satisfying the demands of the AWA. In addition, because the grant money often required matching funds from local government or private donors, zoos became practiced in "assembling" funding packages. The experience of seeking funding from external sources, particularly the private sector, would become increasingly important in the long run.

Although antipoverty and public works grants enabled the partial fulfillment of zoos' increasingly professional mission by, for example, hiring educators and rehabilitating facilities, other federal agencies proved even more useful to their professional goals, especially in the long term. One of the more promising sources of federal dollars came from the National Museum Services Board, created under President Carter in 1976 and charged with reviewing and passing funding for museums and zoos. Millions of dollars were to be made available through competitive grants, matching funds, and loans. President Carter nominated Gary Clarke, the director of the Topeka Zoo, to serve as the zoo representative on the board. Not satisfied with a single representative, the AAZPA had lobbied the Secretary of Health, Education, and Welfare to provide zoos with representation on the board equal to that enjoyed by institutions with "nonliving collections." Assisted in part by Whitehurst, the AAZPA sought a broader definition of museum, arguing that zoos' preservation of endangered species should make them eligible for a fuller range of funding provided by the board. The AAZPA presented a convincing case, and, by 1978, nineteen zoos and aquariums had received grants that ranged between $10,000 and $25,000 for "general operating support." Grants typically went to larger institutions such as Chicago's Brookfield Zoo ($25,000), the New England Aquarium Corporation ($25,000), the Cincinnati Zoological Society ($25,000), and the New York Zoological Park ($25,000). Of the nineteen zoos and aquariums that received money, only four were in cities with smaller populations, including the Santa Barbara Zoo;

Central Texas Zoo, now the Cameron Zoo, in Waco, Texas ($10,000); Inland Empire Zoo, in Veradale, Washington ($7,000); and the Utica (New York) Zoo. This pattern was repeated in 1979, when the twenty zoos receiving the highest grant of $25,000 were the largest zoos and aquariums, some of which had received grants the previous year. Only three smaller zoos received grants, including Clarke's own Topeka Zoo ($14,000).[66] Thus, although congressional testimony and animal welfare groups' research indicated that small zoos needed upgrading most desperately, larger zoos were better positioned to take advantage of the museum services funding, which rewarded professionalism more than need.

Not coincidentally, this was also the period when zoos began adopting the language of museums to a greater extent. Although zoos may have resembled museums in their early history because they exhibited one representative of each species much as an art museum exhibited one typical piece of an artist's work, they did not entirely share the language of museums, largely because, as we have seen, they were generally run by political machines rather than by elite professionals. In 1975, for example, one was unlikely to find references to "curators" in the help wanted section of the *AAZPA Newsletter*. More often, one found requests for "keepers" or "husbandrymen."[67] When references to "curator" positions occurred during this period, they were more likely to appear in the "positions wanted" section and to come from recent graduates of programs in zoology looking for professional employment.[68] Moreover, the larger zoos were more likely than the smaller ones to use the language of the museum world, which suggests that they were more sophisticated and better poised to take advantage of forthcoming museum grant funds. The professional image of larger zoos went beyond job titles to include job descriptions. When the Brookfield Zoo advertised for a mammal curator in 1975, it emphasized the curator's "multifold duties," including "management of animal collection, educational and research activities." By the end of the decade, however, even the smaller zoos and aquariums had renamed positions with more professional job titles. For example, in 1982, a zoo in Peoria, Illinois, advertised for a "curator of animals" who would be responsible for the "care and maintenance of variety of animals" as well as "some keeper/maintenance work."[69] Here we see that, even when the actual job descriptions at smaller zoos may have changed very little, the job titles reflected a new professional self-image.

Like the museum grants, money from the Department of Education, the National Endowment for the Humanities (NEH), and the National Endowment for the Arts (NEA) further solidified the professional credentials of zoos. As we have seen, partly under pressure from environmentalists, animal protectionists, and the AAZPA to justify their holding of endangered animals, zoos had begun to articulate an "education" mission and provide the personnel and facilities to fulfill it. At least in part, federal grants helped both establish education programs and give them the badge of credibility that accompanied federal recognition in this area. In 1978, the NEH made Timothy Anderson, the director of the Franklin Park Zoo, a consultant to the organization to provide it with a "better understanding of the use of the NEH in zoos and aquariums." Among Anderson's duties, he was to attend "AAZPA regional workshops to explain the Endowment's interest in funding certain programs of zoos and aquariums." Another of his tasks was to create a workshop titled "Expanding the Definition of Education in Zoos and Aquariums." Held over several days, this workshop attracted consultants in the areas of "museum education, evaluation, and the humanities" and seemed clearly designed to familiarize prospective grant evaluators with zoos' educational mission. In fact, zoos successfully competed for NEH and even NEA funds. For example, the Shedd Aquarium, in Chicago, received an NEH grant for a project entitled "Lake Michigan and the Humanities: A Humanities Approach to Understanding the Aquatic Environment."[70] This was only one of many federal and state education grants awarded to zoos from the mid-1970s forward.[71]

Education grants were similar to the museum services grants in that they often went to the larger zoos and that the total number of zoos receiving grants was sometimes small. Such grants were, however, of symbolic importance to the zoo profession. The AAZPA, when speaking of its members, tended to blur distinctions between large and small institutions. The professional nature of zoos was revealed not in the total number of zoos that received federal grants, but in the fact that several score zoos *consistently* won such grants and that the smaller zoos were constantly *encouraged* to apply for such grants. Grant awards constituted evidence that zoos were achieving their lofty education and research missions.

Education grants aside, by the late 1970s, other federal sources of funding were becoming endangered and even threatened with extinction. The hardships posed by a weak economy and a more conservative political environment were highlighted for AAZPA members in the June 1979 *AAZPA Newsletter*. In a front-page letter, Joyce Gardella, chair of the public relations committee, reminded members that zoos "live with—and on—the tide of attendance and tax support" and that, in times of economic recession and a "post-Watergate, Proposition 13 era of accountability," zoos would need to cultivate public support assiduously. Gardella urged members to take advantage of the power of the media to "hone the image of zoos and aquariums as a cohesive unit" so that, rather than just being "one more supplicant for the federal ear," zoos could compete "successfully in the federal realm."[72]

The grant money that became available in the second half of the 1970s proved crucial to zoos in several respects. First, federal grants allowed zoos to make some of the renovations and investments in personnel that local governments could not afford. Equally important, zoos polished their professional credentials. To take advantage of some programs, zoos had to emphasize their research and education mission, but federal money in turn helped zoos achieve those missions. Finally, the experience of "selling themselves" to stakeholders beyond local government gave zoos invaluable experience in the art of self-promotion, experience that would prove especially useful when federal funds dried up after Ronald Reagan's election in 1980.

Debating Political Strategy: "Aim for the brain rather than the spleen"

Under Wagner's leadership American zoos expanded their mission and budget. The AAZPA budget also grew, and by 1977 the organization spent more than $50,000 on government services. Accompanying this financial prosperity, however, were internal debates about an appropriate and effective political strategy. Wagner largely bridged the philosophical divisions that had led to ZooAct's creation in 1974, but he moderated continuing discussions about ZooAct's tactics and Steele's image. Those who supported ZooAct and Steele perceived animal protectionists and the federal government as formidable opponents who had to be vigorously resisted. They believed that the AAZPA should, to a large extent, focus on defending zoos' interests, and they welcomed the aggressive posturing brought by Steele and ZooAct. A vocal minority of critics appreciated the necessity of Steele's lobbying activity but preferred for the AAZPA to be represented by someone with real conservation credentials who would be less hostile to the animal protectionists. Those in this camp feared that ZooAct members exaggerated the dangers of antizoo groups, preventing the AAZPA from building meaningful alliances with the protectionist community that would benefit both zoos and animals. Wagner stood between the two camps, extolling Steele's service to the AAZPA even as he pushed for zoos to become energetic conservation institutions.

ZooAct brought financial muscle to the zoo community's political activity. In 1975, when the AAZPA's total budget was less than $135,000, ZooAct collected $20,000 from institutional membership dues and nearly $100,000 in donations. It liked to point out that the "antizoo organizations" raised at least $40 million a year. This figure, however, was greatly inflated by defining as "antizoo" organizations primarily concerned with land conservation, including the Sierra Club, the National Parks and Conservation Association, the Environmental Defense Fund, and Friends of the Earth. Steele lumped these groups together with animal protection groups, including the Fund for

Animals, Friends of Animals, Committee for Humane Legislation, United Action for Animals, Society for Animal Protective Legislation, Defenders of Wildlife, the Animal Protection Institute, Project Monitor, and the HSUS. All of these groups, he argued, were antizoo, implying that some of them worked together in well-orchestrated campaigns "financed by Madison Avenue." ZooAct portrayed itself as leveling the playing field somewhat, distributing brochures with headlines such as "Protestors Demand N.Y. Shut Its Zoos: Your Zoo May be Next," publishing a *ZooAction* newsletter, issuing press releases, mobilizing grassroots campaigns, and taking advantage of Steele's lobbying skills.[73]

ZooAct's publications characterized the antizoo forces as irrational and emotional, but its own rhetoric was unrestrained. A typical *ZooAction* newsletter contained numerous exclamation points, underlined phrases, sarcastic comments, and unflattering descriptions of zoos' opponents. In the January 1976 issue, readers learned about *"the latest humaniac boondoggle,"* a bill authored by two New York congressmen to create a commission to investigate the inhumane treatment of animals. The report noted that in a floor speech about abuses of laboratory animals, Congressman Edward Koch (D-NY) "really called them 'bunny rabbits'!"

This kind of rhetoric dismayed Conway, the Bronx Zoo's director, who wrote Steele to recommend that he employ "the language of reasonableness" in the newsletter and "aim for the brain rather than the spleen." "Educated people," he wrote, "regard ideas couched in unnecessary challenging or prejudiced language with skepticism." Conway appreciated the underlying usefulness of the newsletter's information, but he felt that factual information and logical arguments presented in a reasonable tone would better serve the interests of the zoo community.[74]

Conway's concern was not solely with ZooAct's rhetoric. Their newsletter also misrepresented the activities of animal protection groups, making it difficult for the zoo community to realize that the animal protectionists shared zoos' desire to protect wild habitats and animals. Conway pointed to a *ZooAction* article implying that zoos had played a major role in the population recoveries of a number of species. In fact, Conway observed, in all but two of the cases, "the efforts of protectionists *including NYZS*" had led to the improvements. Conway subtly indicated here that he considered his own zoo's governing organization, the New York Zoological Society, to be one of the animal protection groups that ZooAct often belittled. More to the point, at the 1975 ESA oversight hearings the New York Zoological Society had publicly allied itself with Defenders of Wildlife, the Sierra Club, and the Animal Protection Institute of America in urging Interior to speed up the implementation of the ESA, particularly the listing of endangered species.[75]

ZooAct's failure to recognize the common goals of animal protection groups, land conservation groups, and zoos also bothered some other AAZPA members, particularly those who had strong philosophical commitments to both kinds of preservation and zoos. The Bronx Zoo's Wayne King, who had played significant roles in guiding the AAZPA's conservation policy toward greater protections for wild animals, reacted negatively when a 1976 ZooAct presentation to zoo directors emphasized that several major conservation organizations had actively "opposed legislative positions supported by the zoological community." In response, King circulated a letter "highly critical of George Steele." Conway insisted that King's letter was written in a private capacity, but he shared King's view that zoos should "do everything possible to secure the favorable support of the legitimate conservation and environmental community." In the ZooAct board's discussion of this dispute, Steele recommended that the zoo community "go on the offensive" to educate zoo supporters about the antizoo positions taken by leadership of conservation organizations to which they might belong. An "aggressive campaign," Steele believed, might "persuade" the conservation organizations to support legislation favorable to zoos. In the short term, ZooAct's board members opted for a more moderate program to find, as Paul Chaffee put it, "common ground" with these organizations.[76] Chaffee was a veterinarian who set up his own animal hospital in California and eventually became director of the Fresno Zoo, which was later renamed the Chaffee Zoological Gardens in his honor.

In 1977, this issue spilled over into debate about a dues restructuring plan involving the SEIF, which paid for Steele's services. Set up in 1976 to receive voluntary donations, the SEIF had been funded almost entirely by a handful of large institutions, and in mid-1977 was far short of its $50,000 target. The AAZPA board called a special meeting of zoo directors to explore the possibility of spreading the fund's cost equitably among all member institutions through a restructuring of dues. The eighty directors present voted overwhelmingly to support a restructuring, but the meeting highlighted several concerns about Steele, specifically about whether he "perform[s] in the best interest of the organization." President William Meeker's query on this point generated numerous statements of appreciation for the AAZPA's legislative activity in general and for Steele in particular.[77] For the time being, the question of Steele's service appeared to have been settled, but questions about the AAZPA's political strategy continued to surface.

While Steele's critics accused him of being too hostile toward the environmentalists, his monthly billing details indicate that in reality he spent a great deal of time talking with the very people who had so often criticized zoos. In June 1979, for example, Steele spoke several times with Tom Garrett (Defenders of Wildlife) and Christine Stevens (Society for Animal Protective Legislation).[78] Steele was, as his defenders often repeated, a "professional." Garrett could accuse him during a congressional hearing of crying fake tears for zoos, Steele could call Garrett a humaniac, and the two could later talk privately by phone for an hour. His supporters credited him with establishing a "working relationship with environmental and humane organizations" and appreciated the fact that he had such extensive government contacts.[79]

That same professionalism disqualified Steele in the minds of others. George Rabb, one of the few Ph.D.s among zoo directors, wanted the AAZPA to be represented by someone "able to argue from conviction, and not simply from prowess in rhetoric." Rabb favored someone like Wagner, "a person articulate, informed and committed to the zoo and aquarium profession." This person would need to present the zoo community as composed of "institutions and persons with broad public responsibilities and functions." In a hint at the nature of those functions, he suggested that such a person might be "found among the environmentally aware young crop of lawyers."[80] Rabb's emphasis on environmentalism and public responsibilities aligned him with Conway, King, and the other AAZPA members who hoped zoos would become more vocal conservation advocates.

But conservation legislation sometimes threatened zoos' financial interests, as in 1979, when Steele testified against a bill to create a Channel Islands National Park in California on the grounds that the islands were "one of the best areas for the collection of several important marine mammal species." In the same year, Meritt, as the AAZPA's wildlife conservation and management chair, protested Steele's successful "behind the scene" efforts to prevent the United States from signing a migratory species convention with twenty-two other nations. In response, Wagner argued that as a "professional lobbyist," not a scientist, Steele acted appropriately in defending the AAZPA from a "nightmarish regulatory scheme" that might have resulted from the treaty. At the same time, Wagner affirmed the scientific credibility of Meritt's conclusion that the treaty would provide important protection to many animal species. From Wagner's perspective, however, the AAZPA was caught between a scientific ideal and a political reality, forced to "take it on the chin again" from one side or the other.[81]

The AAZPA retained Steele's services throughout the 1970s, but persistent questions about the conduct of ZooAct and Steele revealed that the organization was not entirely unified in its approach to politics. Confronting doubts about whether zoos had *really* faced "serious legislative threats" in the mid-1970s, Wagner adamantly insisted that such threats had been adverted only because he and Steele had worked so assiduously on behalf of zoos' interests. In his opinion, "had this Association been represented during 1977 as it was prior to 1975, we would have been in serious trouble."[82]

"A Long Way to Go"

Wagner characterized the 1970s as "years of trials and tribulations" during which the AAZPA had increased its political skills and power. Although Congress failed to provide zoos with the financial support that seemed so tantalizingly close in 1974, neither did it take away their freedom to govern themselves. A string of federal zoo regulation bills had died in committee, and by 1976 the AAZPA perceived that such bills had little chance of success. The number of accredited zoos grew from three in 1975 to thirty-three by 1979.[83] This steady growth gave the AAZPA greater credibility as an interest group that did, in fact, accredit and police its members in ways that zoo critics wanted the federal government to do. As two AAZPA leaders remarked proudly in 1977, "We have established much needed credibility with Congressional committees" and "come a long way toward winning the battle for survival in the legislative morass." By 1979, the legislation committee could report a "more positive attitude" in the regulatory arena as well. Despite these gains, zoos had become deeply and inextricably involved with the federal government during the decade, and when the AAZPA compiled a guide to federal regulations affecting zoos, members received a 750-page document weighing eight pounds.[84]

Complicating matters, how best to maintain credibility with the government and conservation organizations had became a somewhat divisive issue among AAZPA members. Those like Conway and Meritt, who fundamentally supported conservation legislation, grew increasingly uneasy with the relationship between the AAZPA and Steele. These members preferred a political strategy in which the AAZPA examined government policy and worked with government officials as scientific peers, much as Meritt had done in the 1978 wildlife management conference. They contrasted this approach to the insider lobbying style at which Steele excelled. This dispute between Steele's critics and supporters reflected a deeper struggle over what kind of organization the AAZPA was becoming: a scientifically oriented conservation organization or an interest group that represented all zoos, including more commercial ones with fewer qualms about species preservation and animal protection. Wagner's job was to assure his members that the AAZPA could be both.

In the context of these developments, AAZPA president Ed Maruska summed up the organization's political future in a 1979 letter to members: "We as an association, have an awesome power structure to create and support strong conservation legislation. We are a power to be recognized and should never permit our role to be overlooked or ignored. However, we still have a long way to go to organize this power structure. We must use our vast resources of influential support, particularly in the legislative areas that continue to hamper our ability to operate with maximum efficiency in the area of captive wildlife propagation."[85] Blakely's apocalyptic warnings had given way to cautious optimism. Maruska's caution, however, turned out to be prescient. Although zoos had scored significant legislative, regulatory, economic, and public relations victories during the AAZPA's first decade of independence, they were about to face daunting challenges to both their economic stability and their legal right to collect, care for, and exhibit wild animals.

FOUR

Species Preservation

"The Shimmering Robe of Light and Goodness"

In the early 1980s Tom Foose, the American Association of Zoological Parks and Aquariums's conservation coordinator, formally announced the implementation of Species Survival Plans (SSPs) designed to help preserve disappearing wild animals. SSPs would "develop scientific and cooperative programs to propagate and preserve endangered species in captivity through professional management." Key to the SSPs was the control of breeding by "specialists" responsible for each species listed. The committee developing the SSP concept emphasized that management and breeding decisions by SSP coordinators should be considered "binding" on all zoos that held the affected species. In fact, the first name for these plans—the master breeding plan—clearly captured the AAZPA's desire to centralize control of breeding.[1]

Centralized breeding became the AAZPA's flagship conservation initiative in the 1980s. The number of animals with an SSP grew quickly, to forty-six by 1988, a trend that promised to alleviate one set of problems faced by zoos—the diminishing availability of wild animals and the burden of obtaining import permits for traffic in species that were not yet self-sustaining in captivity. SSPs also appeared to provide the zoo community with a noble purpose—zoos would serve as the collective ark on which endangered species would be preserved and from which they would be eventually reintroduced into the wild. To reinforce this mission, the AAZPA embarked upon a more aggressive program to support environmental legislation, particularly that which focused on habitat preservation.

Unfortunately for zoos and zoo animals, SSPs exacerbated an already existing surplus animal problem by encouraging breeding and creating a new class of genetically unusable animals. Quite simply, not all captive animals in a given species had the correct genes. These unfortunate animals could not be bred, but took up valuable zoo space and effectively weakened the survival of the species. The SSPs, coupled with the zoological profession's increased ability to keep animals alive for long periods of time, led zoos directly into controversy over how to dispose of animals they no longer needed. Some animal welfare groups cooperated with the AAZPA on issues such as animal auctions and roadside zoos and even supported the humane euthanasia of surplus animals, but others perceived SSPs as "self-survival plans" that benefited zoos at the expense of animals. Complicating the issue, AAZPA members had difficulty agreeing upon a policy that would clarify their responsibility to the welfare of surplus animals.

The AAZPA's conservation mission was also tested by institutions' need to raise revenues through the display of charismatic animals that were in short supply. Sea World's application to capture killer whales drew fierce protests from animal protectionists, while the Toledo Zoo's short-term display of two giant pandas led the AAZPA and World Wildlife Fund (WWF) to sue the Department of the Interior for violating the Endangered Species Act (ESA) permitting process. In both situations, the AAZPA leadership sought to preserve its conservation image, even as it recognized that in the long term zoos would require animals such as killer whales and giant pandas to subsidize their operating expenses. Making the fine distinctions between the legitimate display and the commercial exploitation of wild animals became more difficult as animal protectionists zeroed in on the embarrassing riches resulting from successful breeding programs.

Species Preservation in the Zoo and Beyond

It was clear to Robert Wagner, William Conway, and other members of the AAZPA board that zoos needed a rigorous, centrally controlled breeding program that could ensure the establishment of a reliable long-term supply of endangered animals for zoos. The resulting SSPs promised to satisfy the government's concerns about traffic in endangered species and would even create surplus populations for reintroduction programs required by the U.S. Fish and Wildlife Service (FWS). True to its conservation mission, the AAZPA simultaneously supported legislative efforts to protect the habitats in which wild animals lived.

Zoos embarked on their SSPs in the context of broad international support from environmentalist groups for such approaches to preserving endangered species. The U.S. Man and the Biosphere Program (a committee of the U.S. National Commission for the United Nations Educational, Scientific, and Cultural Organization) invited the AAZPA to cosponsor the 1982 International Symposium and Workshop on the Applications of Genetics to the Management of Wild Plant and Animal Populations. Symposium cosponsors ranged from the FWS to the Sierra Club. In 1983 the AAZPA was invited to join the American Committee for International Conservation, a national committee of the International Union for Conservation of Nature and Natural Resources (IUCN) focused on biodiversity issues.[2] Conservation-focused nongovernmental organizations such as the Foundation to Save African Endangered Wildlife—known as SAVE—willingly helped export endangered animals to American zoos in exchange for substantial donations that would assist conservation efforts abroad.[3] This period also saw the publication of books extolling the role of zoos as wildlife sanctuaries or "arks" in the face of massive extinctions. These books raised the possibility that zoo-bred animals could one day be reintroduced to the wild.[4]

Some AAZPA members focused optimistically on the possibility that SSP's careful management policies would address the undesirable public relations disasters caused by euthanasia because they would "minimize the production of surplus," though Foose cautioned that euthanasia would be "an inevitable method of regulation" for a long time to come because zoos were now quite good at breeding many animals and keeping them alive for a record amount of time. At the same time, zoos were less successful in developing birth control methods that might prevent unwanted animals. The result was increasingly large numbers of animals that needed homes.[5] Convinced nonetheless of SSPs' public relations value and their appeal to the larger environmentalist movement, the AAZPA's executive board moved forward with a plan to get national media coverage of the new program. Soon the board was arguing about the importance of publicity about SSPs in "minimizing the negative impact of our critics."[6]

Sometimes the AAZPA developed SSPs at the request of federal agencies. The ESA required the FWS to develop and implement recovery plans for native endangered species. The FWS, however, lacked adequate human and financial resources for implementing this part of the act. As a result, the agency turned to zoos to breed certain endangered animals for reintroduction. For example, the FWS provided the "impetus on the establishment of studbooks for the Red Wolf and the Mexican Wolf," and the AAZPA board sensed the potential for additional "federal support for conservation programs" (i.e., breeding programs) that would help the government implement its reintroduction goals. In the case of the Mexican wolf, the FWS appointed a "Mexican Wolf Recovery Team" to develop a program that involved capturing some wolves from Mexico to breed in American zoos; and within a few years the agency formally requested that the AAZPA establish SSPs for the Red and Mexican wolves.[7] The Red wolf received an SSP in 1984, and five years later, in the first formal cooperative breeding and reintroduction agreement, the SSP and a FWS recovery plan were combined. Similarly, the U.S.

Department of the Interior determined to capture the last remaining wild California Condors and put them into zoo propagation programs, a decision that caused outrage and a legal battle from some environmentalists. Interior also requested captive propagation programs for birds from Guam and Hawaii.[8]

These cooperative breeding programs effectively placed zoos under the protective shield of the federal government, as indicated by a FWS statement in 1992 that the "[s]ervice welcomes the development" of captive-breeding programs such as "Species Survival Plans of the AAZPA" because they create "a reservoir of animals that can be drawn upon to reestablish or augment wild populations." Moreover, the Office of Scientific Authority (OSA) used the AAZPA as a quasi-public body for exporting endangered captive animals. The OSA required that zoos send their exporting permits to the AAZPA's Wildlife Conservation and Management Committee before sending them on to the government when captive animals were so endangered in the wild that it was necessary to increase them within captivity. The SSPs had become so vital that the government effectively made the AAZPA a clearinghouse for all zoo transactions involving endangered wild animals.[9]

The development of close ties between the FWS's recovery plans and zoos' SSPs helped the FWS defend itself against congressional critics who believed that the agency was doing too little to fulfill the ESA's promise to facilitate the recovery of endangered species. When congressmen accused the FWS of using the ESA simply to obstruct the use and development of private and public lands, the FWS and its allies rattled off a list of species that had been successfully reintroduced. A 1989 General Accounting Office report criticized the FWS on these recovery plans, arguing that resources were focused on species with "high public appeal." Aside from the questionable assertion that Americans feel particularly close to condors—a very large American vulture—the report's authors also chose to ignore the political reality that the FWS had to focus on endangered animals that members of Congress found appealing, because the legislature controlled the agency's budgets.[10]

While congressmen wanted to focus on the charismatic megafauna, zoo professionals were well aware that many wild animals were threatened in the context of habitats and that governments were doing too little to protect those habitats. During the 1980s, the AAZPA gathered with environmentalists in several legislative attempts to create a more visionary federal policy toward ecosystem health. In its first significant foray into habitat conservation, the AAZPA worked with environmentalists to support the creation of an Alaskan wildlife preserve in 1980. Following that success, the AAZPA board provided "enthusiastic" support for a decade-long struggle to develop a bill to protect tropical forests. AAZPA members were urged to support this legislation, because it would give developing countries the ability to exchange their debt for in situ conservation programs. A 1988 omnibus spending bill did include a limited reference to the bill's "debt for conservation" recommendation, but the bill was not fully enacted until 1998.[11]

AAZPA testimony relating to one biodiversity proposal during this period illustrates the combination of self-interest and genuine commitment that the zoo community brought to such issues, as well as the profound way in which captive propagation programs influenced zoos' perspective on much conservation-related legislation. Representative James Scheuer (D-NY), chairman of the House Subcommittee on Natural Resources, Agricultural Research, and Environment, held several hearings during the late 1980s on different versions of his proposed National Biological Diversity Conservation and Environmental Research Act.[12] Scheuer, who had a personal interest in rhinos, had been convinced by previous hearings that "the world's biological diversity, its rich genetic material, its species, its ecosystem are becoming a silent victim to relentless development." To prevent that development and "slow the destruction," he proposed a government regulatory "system whereby science, planning, and coordination are used to prevent losses of biological diversity and to avoid the need for costly last-ditch efforts such as the endangered species act." In practice, this system would establish a national center for biological diversity to collect data and make recommendations that would then be made into coordinated policy by relevant federal agencies; in addition, the legislation would have required environmental impact reports (as required under the National Environmental Policy Act [NEPA]) to consider impacts on biological diversity.[13] In essence, Scheuer sought a preventative approach to conservation, something the AAZPA had advocated in the late 1960s, when it complained that one of the central weaknesses of proposed endangered species legislation was that it had no mechanisms for protecting species *before* they became endangered.[14] The administration and affected government agencies such as the FWS and Forest Service joined with industry interests (e.g., the National Forest Products Association and the Wisconsin Paper Council) to oppose Scheuer's bill on the grounds that government, the private sector, and the nonprofit community were already pursuing diverse and creative biodiversity initiatives and that additional government regulation would only slow this progress and add unnecessary costs to government.

The AAZPA, no stranger to the slowness of government action, nevertheless urged its members to lobby for Scheuer's biodiversity legislation and provided written testimony attesting to its benefits. The AAZPA's remarks revealed both self-promotion and a philosophical interest in biodiversity. First, zoos' conservation mission, their scientific expertise, and their breeding and educational programs were stressed as evidence of their commitment to biodiversity issues. On this front, the AAZPA desired language in the bill extolling the benefits of linking in situ (species preservation in the wild) and ex situ (breeding programs at zoos) conservation efforts. They linked the two preservation efforts by arguing that the "management of captive specimens is an integral part of conservation strategy," protecting "both habitat and species." This self-promotional positioning is, of course, typical of all groups testifying at such hearings, but it is worth noting the extent to which the AAZPA tried to present its SSPs as evidence of zoos' commitment to preserving biodiversity. Second, perhaps fishing for funding, the AAZPA also supported the Defenders of Wildlife in their view that Congress should budget $50 million to $100 million (rather than $5 million) in grants for the initiative. Finally, the AAZPA requested language giving one of its members a place on the scientific advisory committee, which it also believed should have greater authority to develop policy, because zoo scientists would have extensive "experience in linking in situ and ex situ conservation approaches."

To a limited extent, zoos would have benefited from these modifications, but other suggestions spoke to a genuine commitment to finding political solutions to environmental problems. The AAZPA urged Scheuer to broaden his original bill to take into account global actions. Specifically, "Federal agency actions in foreign countries must be consistent with the goal of conservation of biological diversity," because some of the more severe habitat destruction was happening outside the United States.[15] Although Scheuer's bill never got beyond his subcommittee hearings, the AAZPA took some satisfaction in that the congressman rewrote the bill to provide for a federal strategy that included "maintenance of gene pools through combination of in situ and ex situ techniques."[16] Thus we see once again how the AAZPA's captive propagation programs received governmental approval and legitimacy.

The irony of the AAZPA's species preservation work was that it actually exacerbated an existing surplus zoo animal problem. Zoo professionals had been quite successful for a long time in breeding some animals, such as tigers, that were rapidly filling up the exhibit space at American zoos. Although individual zoo directors might encourage species preservation in an animal's native land, they could not require it because they were not citizens of that country. As wild populations declined, zoo professionals felt pressure to increase the number of captive animals, both for display purposes and to create a reservoir of animals for possible future reintroductions to the wild. The result was that AAZPA zoos began to have more animals than they collectively had space to exhibit. Some zoos solved this problem by disposing of surplus animals through a dealer network by which they might eventually end up as pets, in roadside zoos, or even as game at hunting ranches. Because intrastate transfers of endangered species were not regulated by the ESA (only transactions across state or national borders required a permit), even endangered animals could be sold into less-than-humane situations. Species with an SSP enjoyed somewhat greater protection, but individual animals in a given program could be deemed surplus if their genetic profile did not fit the reproductive needs of the larger population or if they were too old to reproduce. This effect of SSPs did not show up right away, but the broader surplus animal problem became an area of concern in the early 1980s and drew the AAZPA into an extended conversation with the Human Society of the United States (HSUS) and Defenders of Wildlife (Defenders).

Roadside Zoos and Surplus Animals

While AAZPA was developing its SSPs in the early 1980s, a complicated relationship among members of the HSUS, Defenders, and the AAZPA emerged over the issue of roadside zoos and surplus animals. Both the HSUS and Defenders followed up on their earlier research into roadside zoos by pushing for state-level legislation aimed at regulating and—they hoped—abolishing them. They also worked with zoo professionals through a special AAZPA roadside zoo committee established in 1980 to investigate the problem. Representatives from the HSUS and Defenders praised AAZPA zoos but also urged them to reform their surplus animal disposal practices. For their part, AAZPA members found it difficult to agree upon a uniform policy governing surplus animals, and the AAZPA board's tentative steps to formulate guidelines and sanctions on the issue were met with resistance.

The AAZPA had actually begun to explore the surplus problem in the late 1970s, at Wagner's urging. Wagner wrote Steve Graham (then director of the Salisbury Zoo, in Maryland) in late 1976 to say that, even though he knew many members would object, the "Association must propose a policy on the distribution of surplus animals." In response to Wagner's request, a special surplus animal committee concluded in 1978 that "the zoo surplus problem is an evidence of [breeding] success" but created ethical problems. The board accepted most of the committee's recommended guidelines "for the disposition of surplus animals," but those guidelines were fairly general. The only "mandatory restrictions" on disposal practices were those specified by state or federal laws. Otherwise, the guidelines were more procedural than substantive, requiring, for example, that disposals (including euthanasia) were made in accordance with a zoo's internally approved policies and that zoos concern themselves with the "best interests" of the individual animals, the species, and the public. Thus, the committee identified a problem, but the board took only the most timid steps to address it. Moreover, the committee invited comments from the HSUS's Sue

Pressman, but did not include her as a member. If the AAZPA hoped to gain public approval of zoos' disposal practices, the board would need to include welfare organizations in the creation of a surplus policy.[17]

By 1980 the place of roadside zoos in the surplus problem had become apparent to AAZPA leaders and animal welfare organizations. For example, Defenders gave East Coast representative Teresa Nelson, located in Boston, the job of pursuing state regulations aimed at curtailing the freedom of roadside zoos. Nelson pushed Maine's commissioner of inland fisheries and wildlife to augment the Animal Welfare Act (AWA) by drafting state regulations specifically aimed at roadside zoos that would cover birds and cold-blooded animals, expand the minimum cage size requirements, create a method for confiscating animals from roadside owners who did not sell to breeders or charge admission, and require that all cages have locks to protect the animals from vandals entering the animals' protected space. Nelson's suggestions were particularly aimed at roadside zoos owned by shopkeepers who wanted to draw business by placing animals outside their establishments. Nelson argued that small private zoos in Maine could "not compete with public facilities designed to teach, conserve and research," noting that they lacked state support that would enable them to pursue these endeavors. And, as was typical of Defenders' arguments, she stressed that conditions in Maine's "menageries" were "atrocious for the animals" and "for the visiting public." She cited, among other abuses, "animals dying of pneumonia on a fairly regular basis" and the sale of surplus exotic animals such as parrots by menageries to the general public. She noted that Defenders was "not anti-zoo," but rather "pro-captive animal." This perspective led Defenders to conclude that roadside zoos should be "upgraded or eliminated" because they were bad for the animals and were "not true examples of wildlife."[18] Nelson kept Wagner informed of her activities and their results, clearly indicating her organization's desire to work with the AAZPA.[19]

Similarly, Sue Pressmen, the HSUS's director of captive wildlife protection, was pushing the AAZPA into focusing on roadside zoos by leveling the charge that roadside zoos existed in such abundance because of surplus animals from established zoos. In 1980 she sent Graham, now the director of the Detroit Zoo, a copy of an article from *Animals,* the Massachusetts Society for the Prevention of Cruelty to Animals magazine, in which she revealed her "private thesis" that "road size zoos would go out of business without the support of bonafide zoos" because they acquired their animals from the surpluses created by zoos' "poorly managed breeding programs." She insisted that the animal welfare community had to "light a fire under the AAZPA to stop supporting the roadside zoos." And she penned a note in the margins of a copy of the article that she sent to Graham asking whether this "inspired him." Evidently Graham had already been inspired because Paul Chaffee, then president of the AAZPA, had recently formed a "roadside zoo committee" by way of memo to some key zoo directors, appointed Graham chair, and asked the committee to examine the size and nature of three interrelated problems: roadside zoos, surplus animals, and animal auctions. The committee included members of both Defenders and the HSUS in a joint effort by the AAZPA and animal welfare groups to combat these problems.[20]

Significantly, the committee agreed with Pressman's argument that roadside zoos were stocking their cages with surplus animals from larger zoos, including AAZPA member institutions. The committee began to collect evidence of the relationship between legitimate and roadside zoos, including hand-printed, poorly written notes from tiny U.S. Department of Agriculture–approved animal exhibitors who offered to take "any babies" deemed surplus. Such crude letters vividly demonstrated the huge gap between professionally managed AAZPA institutions and the hundreds of other places that called themselves zoos. The committee theorized that, if the AAZPA could halt the transfer of animals from legitimate zoos to roadside zoos, it would, as Pressman hoped, eliminate roadside zoos "in five years." Fearing the increased scrutiny of animal welfare activists, Graham demanded that the AAZPA take responsibility for zoos' surplus animal problems "before humane organizations" did it for them. Graham's committee identified exotic animal auctions as one of the main vehicles through which roadside zoos (and sometimes other kinds of zoos) obtained animals and recommended that the board focus on this area.[21]

The AAZPA executive board shared Graham's concern, but it took no new action. Instead, at its March 1981 meeting, it voted unanimously to reaffirm a portion of the existing code of professional ethics that required members to "make every effort" to prevent surplus animals from ending up in the hands of unqualified caretakers. The board indicated that selling animals at auction or giving animals to dealers who sold at auctions would make it difficult to abide by this code, but it stopped short of declaring such actions an outright violation.[22] In short, the board members uniformly disapproved of auctions but were reluctant to translate their disapproval into formal policy. The 1981 statement of position simply bought the board time to develop some kind of consensus among the general membership.

Theodore Reed articulated the antiauction position held by directors from the AAZPA's elite zoos. Reed admitted that he had "many prejudices and negative opinions" about auctions from working in domestic livestock auction yards in Oregon in his early veterinary career. He noted that, because auctions were typically held far away from zoos, they caused a "tremendous amount of stress" for the animals transported to unfamiliar surroundings.

They were also prime breeding grounds for disease as sellers brought together animals from all over the country. He characterized them as a "crass business" because the sellers had "no concern for where the animals come from; where they go; or how the animals are treated." Instead, the auctions had the feel of an "Old Yankee trader seeing who can put over the best deal" to get rid of "bummers, shy breeders, the lame, halt, and blind." In Reed's opinion, a zoo director had a "moral obligation" to his animals and ought to care about where his "animals or children" went after they left the protection of the zoo.

Reed acknowledged the difficulties of creating a publicly acceptable animal disposal policy that fit with the current scientific limitations of zoos' animal management programs. In theory, he preferred that zoo animals be taken to a slaughterhouse, where they would be "humanely killed and sold for food rather than sent out to be abused by anyone with a buck." Zoo directors who used the auction houses instead were unwilling to honestly face up to the problem that they have with surplus animals. An alternative, he proposed, might be to sell some animals directly to the public for the private pet trade. Reed offered no easy solution, but he summarized by arguing that the AAZPA was becoming a "more professional, idealistic" organization devoted to the principals of "genetic integrity" and the "preservation of the species." Gone were those "happy, sloppy, independent days" when they were not regulated by the federal government or watched by animal protection groups. Now they must "cooperate, plan, and establish standards and rules" and then "live up to them."[23]

The surplus issue placed Wagner in the position of trying to hold his membership together and help his executive board evaluate who should be disciplined for potential ethics violations arising from the sale of animals at auction or to inappropriate parties. His response to the zoo directors' dislike of auctions indicated the stresses of this balancing act. Some directors wanted more aggressive ethics policing that would have resulted in temporary suspension or termination of institutional membership if the ethics board (an elected subcommittee) found a zoo director guilty of a violation. Others favored an official reprimand from the board, an action that would express disapproval without imposing any real penalty. Without any real agreement, the AAZPA's actions against members proceeded haphazardly and, in Wagner's opinion, unfairly. In a late December 1981 discussion of this problem, Wagner described how one member had recently offered ostrich for sale in a Neiman-Marcus Christmas catalogue, whereupon the ethics board investigated the sale and unanimously voted to reprimand the director and the zoo. In contrast, he noted that there were still AAZPA member animal dealers selling animals at auctions who had not been reprimanded, despite three public warnings in 1981 that the board was "unequivocally opposed to such auctions." He worried that the association was not conducting their affairs in "an evenhanded manner" and failed to see that the ostrich seller was any more "guilty than our animal suppliers and other members who offered wildlife for sale at auctions."[24]

Wagner's task was made more difficult by the fact that even the board's simple statement of a surplus philosophy was not universally welcomed by AAZPA members. Several directors expressed their personal dislike for auctions but advised Wagner to restrain from imposing "too many laws," "strict 'rules,'" or "reprimands" on the members, because many "obviously are not listening to us anyway." One director noted that the issue could potentially "split our organization." Another expressed his concern that some members were "'fed up' with the 'dictatorial' edicts emanating from our Board in many areas" and felt they did "not need the AAZPA."[25] External pressure, in the form of negative publicity and criticism from animal welfare organizations, kept the board concerned about the surplus problem, but internal disagreements prevented it from finding a solution that would not threaten the association's unity.

Moreover, Wagner knew about the limitations involved in not selling surplus animals directly to pet owners. The reality that he faced was that there were elite Americans who owned large numbers of exotic animals that could potentially be used in the SSP programs. To ignore these old-style private menageries risked losing both genetic variety and a place to put their surplus animals. A former president of the Houston Zoological Society, Tyson Smith, for example, owned a "2,000 acre ranch" in Texas on which he raised "approximately 340 varieties of hoofed animals, monkeys, and birds." What complicated matters for the AAZPA, however, is that animals in these collections were occasionally hunted. Although Tyson primarily bred the animals and forbade hunting on his ranch, he admitted that he sold older male hoofed animals to hunting ranches where they would be "everlasting" as trophies instead of "dying in the pasture."[26]

To demonstrate the zoo community's concern, the board kept Graham's committee on the surplus issue. Through 1982 his committee (renamed the Animal Welfare Committee) continued to investigate exotic animal auctions for the purpose of enlightening the AAZPA board on the issue. Graham's 1982 midyear report expressed his revulsion at the "barbaric" conditions he had witnessed at one auction where animals were fed roadkill and housed in tiny crates. He feared that such auctions were a "ticking time bomb," and he implored the board to "beef up" the association's surplus policy. The public relations danger of the auctions had been revealed by HSUS investigations that found AAZPA members participating in such auctions, an action that

Graham personally witnessed and felt should result in ethics violation charges. By 1983, Graham's committee had come to see the truth in Pressman's contention that the auctions were "but the tip of a very large iceberg" of problems relating to the unregulated disposal of surplus animals. As Ronald Blakely observed, it "might seem easier to say 'species first, individuals second,'" but doing so would only "wave a red flag in the faces" of protectionists who cared mainly about the fate of individual animals.[27] Defenders and HSUS appeared to have become reliable allies for the zoo community, but they represented only two organizations.

The AAZPA and its zoos were at the vortex of key conservationist and animal welfare debates about what to do with increasing numbers of animals that were living longer and had little chance of being returned to their native lands. Board members may have disagreed over specific ethics violation questions, but they were united in their desire to protect the zoo community from the negative publicity caused by individual zoos' inhumane disposal of surplus animals. At the same time, members of the board did not want to turn their zoos into large animal nursing homes with collections of aged, dying animals that were never replaced. Animal protection groups talked a lot about species preservation and the welfare of animals, but few of them engaged in any direct protection themselves, because they did not own the very animals that they hoped to preserve. Somehow zoo directors had to translate their own interest in species preservation as well as their critics' into a *practical* program while simultaneously struggling with limited knowledge of animal birth control and avoiding putting themselves out of business with too much negative publicity about animals for which there was no space. As they struggled to deal with their logistical problems of animal preservation both internally and with animal activists, events were unfolding that forced the topics of surplus animals and euthanasia into the public and ironically involved the chair of the roadside zoo study committee, Steve Graham.

Big Fuzzy Cats

In late July 1982, Graham gave an interview to the *Detroit Free Press* in which he mentioned "the possible need for euthanasia of overcrowded zoo animals" for humane reasons. As an example, he noted that the Detroit Zoo exhibited seventy aoudads (a kind of North African sheep) in a space designed for only fifteen. As a result of this interview, "a ripple of disapproval of the idea of euthanasia as a planned part of zoo management policy washed through the community." Doris Applebaum, an AAZPA member and an employee of the Detroit Zoo, described this "ripple" becoming a "tidal wave" when another front-page article in the same newspaper revealed that the zoo planned to euthanize three Siberian tigers—and possibly a fourth. Three of the tigers (Czarina, Nick, and Boris II) were elderly and suffered such ailments as kidney disease and severe hip dysplasia, while the fourth (Anna Scraya) was healthy, but had developed an "aggressive temperament" that Graham worried would threaten the other tigers in the exhibit. At the same time, all four of these tigers (along with four other tigers from the zoo) were part of a nutrition study coconducted by researchers at Michigan State University and were required to undergo several anesthetizations from which the ailing tigers had had difficulty recovering. Thus, to protect the tigers from further suffering, the zoo decided to euthanize them at the project's conclusion by administering an extra dose of anesthesia.[28] This decision prompted a court challenge from animal protectionists and led the zoo community into a broader battle over the allegedly inhumane effects of SSPs.

Graham, who was also the chairman of the AAZPA's animal welfare committee, approached the euthanasia decision in an open manner. Sensing the potentially controversial nature of the situation, he called a "precedent-setting" meeting with his senior staff who, after a "traumatic, difficult, and trying" discussion, reached unanimous agreement on the necessity of euthanasia; this decision was then communicated to other zoo personnel.[29] Graham's best efforts to assure consensus were undone when a presumably "disgruntled zoo employee who disagreed with the Director's management policies" leaked the information to a local reporter whose first article on the tiger euthanasia "placed major emphasis on [Graham's] previously quoted statements about culling."

Graham soon discovered that not just his staff, but also the public, felt they had the right to shape zoo policy. After the story broke, the zoo's phones "started ringing and did not stop for two days," and letters "started pouring in" from around the nation. The AAZPA described the tone of the letters as "unbridled, [and] almost hysterical." The zoo worked vigorously to correct the common misperception that the tigers were healthy (in fact, the one healthy tiger, Anna Scraya, had been granted a reprieve for another research project even before the controversy broke). All letter writers received a personal explanation of the situation from Graham, and some even "wrote back, rather sheepishly apologizing for jumping to a conclusion on the basis of incomplete knowledge."[30]

One particularly upset citizen, Krescentia Doppelberger, a fifty-nine-year-old Detroit resident who had immigrated from Bavaria, filed a lawsuit against the City of Detroit (the zoo's governing authority), contending that the zoo's euthanasia decision had been "arbitrary and capricious and malicious" and seeking a court order to prevent the killings and monetary damages for emotional suffering. Explaining her reason for filing the suit, she said in her broken English, "everybody just complain but never anybody do anything." She "just didn't agree" with the euthanasia decision and "as a taxpayer" felt that she owned the animals, because the Detroit Zoo was a public institution; therefore, she should have the right to question policies affecting the animals. She and her husband favored euthanasia if the animals were truly ill: "if they are in this much pain as they say . . . we're the last ones to prolong the animals' sufferings." She simply wanted to make sure that "what is done is the right thing to do." Her background, according to a local news story, involved reading "*National Geographic* assiduously," sympathizing "with attempts to protect great whales and Alaskan baby seals," and harboring a "changing cast of domestic and wild creatures." When a neighbor once complained about one of Doppleberger's cats wandering into her yard and asked Doppleberger why she needed three cats, she replied, "because there are people like you who don't like them."[31]

The legal proceedings initially went against the zoo. Cleveland Amory—the head of Fund for Animals, a member of the HSUS board of directors, and a well-known conservationist—joined the plaintiff in the suit, claiming that the euthanasia decision would injure him by interfering with his "rights and responsibilities" in "preventing cruelty to animals in the State of Michigan." At the first hearing in the Wayne County Circuit Court, on September 24, 1982, a large crowd heard Amory tell the judge that "animals can take an awful lot of suffering. But when you take a life, you can't bring it back, sir." The presiding judge issued an injunction "ordering the zoo to delay the planned euthanasia until a court hearing could be held" and ruled that the plaintiffs were to "call in some outside veterinarians" to evaluate independently the necessity of euthanasia. In a minor victory for the plaintiffs, the judge also rejected the zoo's request that all expert veterinarians be members of the American Association of Zoo Veterinarians, though he did grant the zoo the right to euthanize the tigers at *any* time if medically necessary, and he barred the media from attending the medical examination.[32]

The question of the plaintiffs' standing occupied the courts throughout the proceedings. Recognizing the danger that an unfavorable outcome might encourage citizens to challenge in court the decision of *any* city department head, the City of Detroit appealed to the Michigan Court of Appeals for a summary judgment. The appeals court declined to rule on the case and sent it back to the circuit court. In a dissent, however, one judge argued that the defendants should be granted summary judgment because experts could reasonably disagree over the necessity of euthanasia. In fact, he opined, "the decision to euthanize the tigers is a *political one* which is not justiciable and plaintiffs' remedy, if any, must come through the political process." This opinion coincided with the city's view that Doppleberger did not have standing just because she was a taxpayer. The city cited other Michigan cases that held "a taxpayer can not assert public rights similar to that enjoyed by the general public." The circuit court judge, Paul Teranes, who was blind and used a guide dog, grappled with the standing question, recognizing it as the core issue to resolve. On November 3, 1982, he ruled that the plaintiffs' mental distress did not give them standing any more than similar distress would give a citizen the right to sue the police over a shooting that did not directly involve that citizen.[33]

The injunction against the zoo was lifted largely on the strength of testimony from expert witnesses who supported the view that euthanasia was humane, necessary, and consistent with professional zoo standards. The zoo could point to an independent twenty-two-member "medical advisory council" that "unanimously supported the zoo's plan for euthanasia." During court proceedings on November 3, 1982, several expert veterinarian witnesses confirmed the zoo's initial claims that one of the tigers was "in so much pain from a dislocated hip that he leans against the wall (of his cage and) uses it like a crutch," that a second was vomiting from kidney damage, and that a third also had "a hip problem and limped." Perhaps more compellingly, Wagner came to articulate the AAZPA position that "there is little doubt that euthanasia is an appropriate management tool—especially when euthanasia is recommended for sick, injured or superannuated zoo specimens." Wagner's testimony affirmed that the Detroit Zoo's euthanasia decision was consistent with the AAZPA's official surplus animal policy. Wagner indicated that, if Graham had sent the tigers to another facility (as the plaintiffs wanted), he might well have been found guilty of an AAZPA's ethics violation, because he would essentially have been "pass[ing] the buck" on making a decision.[34]

To the relief of the zoo community, the court had taken the view expressed by Pressman. Persuaded by both the AAZPA and the medical advisory council's professional recommendations, the circuit court allowed the zoo to euthanize the three sick tigers, and it later lifted the injunction against the fourth tiger. Applebaum noted that "this was the decision that the zoo had hoped for, because it meant that private citizens could not take zoo officials to court simply because they disagreed with the decisions made at the zoo." Repeating Wagner's earlier arguments, the court "made it clear that, as long as zoo officials' decisions were made in a reasonable manner and were not arbitrary or capricious, their professional judgment could not be challenged in court." Applebaum summarized by noting that the "precedent set by this decision is an extremely important one for the cause of professional zoo management."[35]

Although a legal defeat for the animal protectionists, the Detroit Siberian tiger case forced the AAZPA to publicly articulate its animal management policies, policies that it began to reconsider partly in response to this case.

Whereas euthanasia of sick animals was easily justified to the public, the AAZPA also invoked species "management" to suggest that the killing of healthy animals could be necessary. In court, the city's lawyer captured the dilemma that faced zoos: "The goal of preserving endangered species is not that we preserve every individual, but it is to preserve the species as a whole." Wagner made the same point when he revealed that the four tigers could not participate in AAZPA breeding programs because of "genetic problems." In other words, the most valuable animals in the zoo were those that could contribute to the development of a sustainable, captive-bred population.

Wagner and other zoo directors who knew about this case, however, recognized the political pitfalls caused by euthanasia. The zoo's lawyer summed up the problem bluntly: "The only reason we're here is because we have some nice big, fuzzy cats and people get upset. . . . to save one old tiger you are going to take up space that can be used to breed new and healthier tigers." The potential pitfalls of a species management justification for euthanasia were vividly expressed by a correspondent who later asked Wagner, "Do you want to be known as the savior of endangered species or as the Hitler of the animal world who advocates the death of any animal not on your endangered species list with pure proven bloodline!" Wagner realized that since the "fuzzy" tigers were "favorites of citizens that visit the zoo," the situation created a public relations problem of the kind that other zoos might well face. In a later report to AAZPA members, he expressed his concern that "our members carefully consider all of the ramifications of euthanasia" and asked that then President Peter Karsten consider dedicating a session of the 1983 annual conference in Vancouver to "a discussion of euthanasia."[36]

Continuing its cooperative work with the AAZPA's animal welfare committee, the HSUS had taken the Detroit Zoo's side during the trial, arguing that it was more humane for zoos to euthanize surplus animals than to sell them to roadside zoos. Pressman had sent a letter to Judge Teranes on behalf of the HSUS supporting the Detroit Zoo's decision to euthanize the tigers. She implied that, if the judge ruled against the zoo, his decision would call into question "a long accepted veterinary practice" to euthanize animals suffering from untreatable and painful ailments. The judge should see the case in the context of the millions of American families who wished to end the misery of their ailing pets. Although it was unfortunate, zoos had to use euthanasia to deal with "the problem of surplus animals" that could not be given homes in "other acceptable facilities." Pressman elaborated on this latter point in a second letter to a concerned HSUS member. She stressed that Graham's action was "humane," compared to the alternative employed by "too many zoos" that had "taken the expedient, highly unprofessional" route of selling their unwanted animals to "roadside menageries." Moreover, Pressman echoed the AAZPA's position that zoo management decisions should consider not just the welfare of "individual animals," but also "the good of the species." Graham and his counsel tried to use the HSUS's euthanasia position in their defense at one point in the proceedings, eliciting laughter from spectators supporting the Fund for Animals.[37]

Pressman's involvement in the case revealed that the HSUS desired to continue its cooperative relationship with the AAZPA on issues such as roadside zoos. Following the Detroit trial, Pressman asked Wagner to publicize the HSUS position in the *Newsletter.* Recognizing that the zoo community had a "growing sensitivity" to the euthanasia issue, she particularly wanted it made aware of the fact that not all animal welfare organizations opposed euthanizing "ill or surplus animals." Wagner fully realized that Pressman's distinction between ill and surplus animals was a crucial affirmation that the HSUS "totally and completely supports our SSP program and the inevitability of euthanasia." Wagner appreciated Pressman's position, and he encouraged her to get involved in the 1983 AAZPA conference where the euthanasia issue would be addressed.[38]

The 1982 Detroit Zoo tiger case did not put to rest the euthanasia issue. In fact, the zoo's euthanasia three years later of four healthy but genetically unfit Siberian tigers attracted renewed attention to the role of SSPs. The four animals had been excluded from the Siberian Tiger SSP, no AAZPA member wanted such animals, and the several interested nonmember zoos were deemed to have "inadequate housing." Although Graham euthanized the animals before contacting the media, the act generated little public controversy. Protectionists could hardly organize to prevent an act that had already occurred. Nonetheless, the Animal Protection Institute (API), which had opposed the zoo's 1982 euthanasia and had already expressed deep concerns about SSPs, was particularly outraged by the 1985 decision.[39]

After the 1985 euthanasia, the API's president, Belton Mouras, warned Wagner that the SSPs were putting the zoo community on a collision course with animal protectionists. Mouras insisted that the "last thing in the world" he wished to do was get into a "battle with zoos" because he recognized that so many of his members enjoyed them. Privately, however, he questioned why zoos were setting up a system like the one we had for "unwanted dogs and cats" in which we kill them when there are no takers? He demanded that zoos assume a lifetime responsibility for all animals produced in their institution until their health was so impaired that all "humane workers" could agree that they ought to be euthanized. Mouras scorned the "elitist principles" behind a species management justification of euthanasia, and he promised to bring this problem to the attention of the Summit for the Animals (an annual gathering of animal welfare organizations). Forebodingly, he warned that "it will be my proposal that we make the arresting of this misguided policy a goal for all groups generally." To forestall this possibility, Mouras sought to bring the API and other groups into formal discussions with the AAZPA, but it was obvious that Mouras's underlying concern for individual animals could never be made compatible with the AAZPA's species management approach.[40]

What emerged from the Detroit Zoo case was a shifting array of support and criticism from animal protection groups. At the same time that one member of the HSUS board took the Detroit Zoo to court for euthanizing animals, another member wrote a letter to the judge supporting the zoo's euthanasia decision. The API meanwhile, broke ranks with the HSUS and condemned an approach to animal management that involved euthanasia. What all groups shared, however, was the desire to have a say in control over animal welfare decisions at zoos.

For its part, AAZPA leaders discovered that the surplus problem got larger the more they examined it. By 1985 the Animal Welfare Committee was examining not just auctions, but also the use of zoo animals in mall shows, as sports mascots, and in the exotic pet trade. In 1986, the committee began to consider the placement of surplus animals in research laboratories and circuses. Throughout this period, the HSUS continued to cooperate with the committee, but it was becoming clear that the AAZPA membership was so divided over the surplus issues that the board could not realistically expect to impose much discipline on zoos' practices. When the committee surveyed zoo directors about their feelings on some of these issues, it found them evenly divided on whether it was ethical to sell surplus animals to researchers, circuses, or the pet trade. Directors were in agreement on only one point: zoos should not be responsible for their animals beyond the point of initial sale. Graham threw up his hands, wishing the committee's new chairman, Ken Kawata, "all the best luck," but warning that, after ten years of working on the surplus issue, "frustration" was the only "mantle" he could pass on.[41]

As it turned out, citizens such as Doppleberger and animal protection groups found additional reasons to question zoo directors' decisions. While the Detroit tiger case tested the question of what zoos should do about elderly ailing animals in abundance at zoos, it left unanswered the question of what they should do about animals that were few in number in captivity. Should they let the existing killer whales and panda bears die, preferably of old age, in captivity without any attempts to replace them with animals from the wild? And what should they do about endangered species that might actually disappear in the wild? Should zoos import them for breeding purposes or leave them to their small numbers in the wild and, potentially, let them die off? Two court cases in the 1980s forced the AAZPA and member institutions to confront these questions.

Budgies, Killer Whales, and Panda Bears

The zoo community had fought for and largely secured the legal right to take limited numbers of endangered species from the wild. This right, however, was not absolute. Individual permit applications were subject to public comment under the Marine Mammal Protection Act (MMPA), the permitting process itself could be revised, and granted permits could be challenged in court. The AAZPA needed to protect its member zoos against all three kinds of challenges, because most SSPs inevitably required at least the occasional introduction of new genetic material from the wild. At the same time, the AAZPA needed to prevent its members from pursuing their own interests at the expense of the larger zoo community's conservation image. Two cases represented the opposite positions that the AAZPA could take when a member sought to acquire highly visible and controversial animals. In the first, the AAZPA supported Sea World's bid for killer whales, believing that the animal protectionist community's opposition, if successful, could inspire additional restrictions on other species necessary for the SSPs. In the second case, the AAZPA used the courts to challenge a member zoo's acquisition of two giant pandas. Again, the association took a long-term view, realizing that if zoos hoped to exhibit pandas in the future, they needed to conform to a strict importation policy that would support captive propagation efforts. The first case to test whether zoos and aquariums could continue to collect protected wild animals was *Jones v. Gordon* (1986).

In *Jones v. Gordon*, the National Marine Fisheries Service (NMFS) and Sea World appealed a district court order that granted summary judgment in favor of Tim Jones, a tour-boat operator who previously argued that the NMFS permit granted Sea World was invalid. At stake was Sea World's right to capture and exhibit killer whales.[42] In discussing this case with AAZPA members, Wagner warned that, if animal protectionists could prevent the taking of marine mammals, they would "continue to whittle away until zoological facilities will be hard pressed to obtain permission to display budgerigars."[43] As in the Detroit tiger case, the real issue here was one of control. Animal protectionists hoped to use the court system to give the "public" a stronger voice in decisions made by the bureaucracies that regulated zoos. The AAZPA preferred to have its members deal directly with bureaucrats, who, in general, tended to support the zoo community.

In March 1983, Sea World, following MMPA guidelines, applied to the NMFS "for a permit to capture killer whales for the purpose of scientific research and public display," the only accepted taking of this species under the MMPA's general moratorium. Over a five-year period, Sea World desired ten killer whales for permanent display and research in its facilities and as many as ninety for temporary (less than three weeks) study. All whales would be collected off the West Coast of the United States. After receiving the request, the NMFS invited four months of public comment and "received approximately 1,200 comments supporting the application and 1,000 comments opposing part or all of it." Those opposed to the permit persuaded the service to hold a two-day hearing on the application at which a host of pro- and antiaquarium forces turned out to voice their opinions.[44]

Opponents of the permit included Senator Slade Gorton (R-WA), Representative Don Bonker (D-WA), and Representative Norman Dicks (D-WA), as well as Washington State's secretary of state, Ralph Munro, and commissioner of public lands, Brian Boyle. The Washington political opponents stressed their right to have wildlife remain in their state. They argued that Americans who were interested in seeing killer whales should come to the Puget Sound area.

Supporting this delegation's position against taking the killer whales were Greenpeace, the Fund for Animals, the American Cetacean Society, and API. One of the founding members of Greenpeace, Paul Spong, had worked in the Vancouver Public Aquarium, from which he had been fired for claiming that Skana, the captive orca, had "told him" that she "wanted to be set free." Spong and another Greenpeace founder, Robert Hunter, had played a crucial role in saving whales, spearheading Greenpeace's work in stopping Soviet ships from them killing them, for example. Thus, Greenpeace was not anxious to have their work undone by an American institution.

The API, meanwhile, used Greenpeace's legendary direct action against international whaling ships as a threat to try and stop Sea World from pursuing its bid to capture the whales. In a 1983 letter to Sea World's Lanny Cornell, senior vice president of Sea World, Mouras, warned that the "same Greenpeace warriors who chased the Russian whalers" would now set off after Sea World ships trying to catch the killer whales. He reminded Cornell of the public relations disaster that this direct political action would have for Sea World; and he argued that Sea World was failing to set a "decent example" for Japan and the Soviet Union for upcoming whaling negotiations. Finally, Mouras appealed to Cornell's "magnanimous" and "business shrewd" sides to arrange a "crash meeting" to negotiate a compromise which "leading objectors" to Sea World's plan were willing to live with. Besides, he argued, even if Greenpeace never got to the point that it was placing zodiacs (small motorized boats) into the water and chasing after Sea World's ships as it had done to stop Russian whalers, he presciently predicted a "tidal wave" of disapproval once the public heard about Sea World's plan.[45]

The opponents were joined by David Hancocks, then director of Seattle's Woodland Park Zoo. Abandoning solidarity with both zoo directors and the AAZPA, he argued that the history of zoos' attempts to preserve certain species was not particularly successful. Hancocks gave the example of gorillas, reminding the committee that "for every gorilla that reached the zoo, five died in transit." He also rejected the position that aquariums, if allowed to take a few of the whales, would be able to breed meaningful numbers. He speculated that "hundreds of killer whales would have to be taken to achieve a self-sustaining population." Finally, he criticized Sea World's killer whale shows as "a distortion of reality [that portrays] the killer whales as 'mere talented freaks.'"[46]

In contrast, proponents of the permit emphasized the elitist nature of the arguments against capturing the whales. Touching on this theme, conservative columnist George Will wrote an article for *Newsweek* after the regulatory hearings in which he championed the side of Sea World in part on the basis of the number of people who learn about marine mammals by visiting its parks. He particularly stressed what he felt was the economic bias of Munroe's argument by pointing out that the Munroe family had lived in the "Puget Sound area for four generations." Munroe had argued that the citizens of Washington "were tired of California amusement parks taking [their] wildlife down there to die" and wondered why the children of Washington should have to go to Southern California to see their native animals. Will hoped that if Sea World was denied its permit "230 million Americans" would "go to Puget Sound, unfold lawn chairs on Munroe's lawn, ask for iced tea and watercress sandwiches and watch the whales." Missing the subtleties of Will's arguments, which had actually disparaged "Nuke the whales" proponents, Republican Representative Don Young (R-AK) introduced this column on the House floor and submitted it to the *Congressional Record* to express his concern about "environmental extremism."[47]

Showing up at public hearings in favor of the permit were Wagner, Warren Zeiller (general manager of Miami's Seaquaria), and William Braker (director of the Shedd Aquarium).[48] Sea World representatives argued that "75 million people" had seen killer whales at Sea World parks and learned about them firsthand. They reviewed the existing research and argued that these new whales too would be used for research on geographic variations (they planned to take the whales from along the Pacific coast), as well as for information about parasites and bacteria, among other studies. Directly addressing the suggestion by animal welfare activists that Americans should get their information from nature documentaries rather than direct experience, William Braker argued that "the scent of a ferret, the exploratory probing of a baby elephant's trunk, the constrictions of a boa and the inundating splash of a killer whale cannot be experienced through the lens of a camera." He noted that increasing nature shows had not reduced Americans' desires to see real animals. Rather, they had increased it. The problem with animal welfare activists' suggestion that Americans see killer whales by taking whale-watching cruises was that it was an "elitist" approach, because most Americans did not have the money to make these trips. He noted the irony of charges of "commercialism" against Sea World and the lack of similar concerns about whale-watching watching tourist boats. Finally, he stressed that the killer whale taking and research were vital to the zoo community's larger conservation mission: the "captive breeding programs that presently are the only salvation for many endangered and threatened species." Even though killer whales were not endangered, Braker feared that a permit denial would have a chilling effect on other captive propagation programs.[49]

Cornell defended his institution by insisting that "we at Sea World are environmentally concerned citizens," a statement that evidently "drew laughter from the crowd of about 150" people at the hearing. He also shielded Sea World against the accusation that the research on the animals was "just window dressing" to "cover the taking for public display" by countering that Sea World could simply capture the animals for "considerably less" money than they were committing to research the animals.[50]

Following this contentious public comment period, the NMFS service issued a permit to Sea World but restricted the research it could undertake and limited the number of whales to two for exhibition and thirty for research and return to the wild.

These restrictions, however, failed to satisfy animal protection groups or politicians from the state of Washington. A spokesperson for the API publicly condemned the issued permit by arguing that it jeopardized the "five-year moratorium on whale hunting" that was going to start in 1986 and would give "the Soviet Union and Japan a way to wiggle out of it." And Munro charged that Sea World's real purpose in conducting research on the whales was "to [learn] how [to] keep them alive in a swimming pool, not how to let them live in the wild."[51]

Acting on these concerns, Jones "sought declaratory and injunctive relief" against the NMFS in the U.S. District Court for the District of Alaska, alleging that the service violated the 1969 NEPA by issuing the permit "without preparation of an environmental impact statement" for an activity that had potential "environmental consequences" and had caused significant "public controversy." For its part, Sea World contended that they were burdened by two regulations, NEPA and the federal comment requirements that conflicted with one another in terms of time frames, and thus they argued that they did not have to do an environmental impact study. The U.S. Court of Appeals for the Ninth Circuit disagreed, stating that Congress had clearly wanted NEPA to "apply to the fullest extent possible." Although the court acknowledged that rules on environmental impact statements for the public display of marine mammals had been revised by an administrative directive in 1980, even that directive required an environmental assessment in decisions involving "public controversy based on potential environmental consequences," which was clearly the case here. In addition, the court found that there was no time conflict because the NMFS "could withhold publication long enough to comply with any NEPA requirement for preparation of an environmental impact statement."[52] Doing so, however, would delay Sea World's collection of the animals.

Ultimately, the court of appeals ruled that "the district court did not err in deciding that the NFMS has unreasonably decided not to prepare an environmental impact statement," but it declined to say that Sea World *had* to prepare a statement. Instead, it overruled the lower court's order and required instead that the service must "consider the requirements of NEPA and regulations" and provide a reasoned explanation as to why "Sea World should, or should not, have to file a statement."[53]

On the one hand, the animal protection groups received standing in the Sea World case, but they achieved only a partial legal victory, succeeding in forcing the NFMS to reconsider the permit application but failing to get the court to order an environmental impact statement. On the other hand, publicity surrounding the case led some members of Congress to consider a ban on taking killer whales altogether. In the end, political opposition quashed Sea World's permit application, but animal protection groups were unable to restrict significantly the right of aquariums to take other kinds of marine mammals in the long run.

To prevent similar attacks in the future, the AAZPA pursued legislative assistance. Even though the courts rejected the animal protectionists' attempt to use federal policy as a tool to constrain the behavior of aquarium directors, the AAZPA began working to ensure that future modifications to the MMPA would even more explicitly protect zoos' interests. In 1988 the MMPA was amended in several ways that affected aquariums. First, at the request of the AAZPA, Congress created a special permit category to allow takings or importations that promised to "contribute significantly to the survival of the species." In other words, aquariums would now be able to argue that captive breeding of captured endangered animals could promote the survival of the species. For their part, the animal protectionists persuaded Congress to require aquariums seeking permits to "offer a program for education or conservation purposes," though the AAZPA succeeded in getting language recommending its own "professionally recognized standards" for public display. Furthermore, Congress reaffirmed that "it is not the intent of the Act to prohibit the display of marine mammals in zoos, aquaria, or amusement parks that comply with applicable regulations and standards."[54]

After passage of the 1988 MMPA amendments, the AAZPA attended related NMFS hearings, testifying that it was unnecessary to determine an optimum sustainable population for public display permits and—at the NMFS's request—recommending "definitions" for "conservation and education programs" related to marine mammals. In effect, the AAZPA managed to take control of the education and conservation requirement by establishing its own definition. Later the AAZPA and its members lobbied against a proposed amendment that would have allowed state governors to veto public display taking permits issued by the NMFS.[55]

Zoos' legal victories against animal protectionists provided them with some breathing room, but the Detroit Zoo and Sea World cases focused public attention on some unpleasant operational realities that zoos preferred to keep hidden. To a certain extent, however, zoos managed to set the media agenda by publicizing their education and species preservation initiatives. In both cases, moreover, the AAZPA presented a fairly united front against animal protection groups. Neither case, however, allowed the AAZPA to make vividly real its core claim of being a conservationist organization. That opportunity would come a few years later, when the Toledo Zoo decided to import two giant pandas.

The AAZPA had been reluctant in the 1970s to admit more than a handful of for-profit zoos to full membership, partly because the Convention on International Trade in Endangered Species of Wild Fauna and Flora (CITES) had language specifically prohibiting trade in endangered animals "primarily for commercial purposes."[56] This wording was vague enough to be interpreted to mean that a for-profit zoo could not trade in endangered animals. Government regulators, however, accepted the AAZPA argument that if a for-profit zoo had robust education and conservation programs (through, for example, participation in SSPs), it would not be displaying endangered animals for *primarily* commercial purposes. A trickier situation arose when a nonprofit zoo imported an endangered animal mainly to increase revenues, and without AAZPA support. This happened in 1988 when the Toledo Zoo imported two giant pandas from the People's Republic of China (PRC) for the purpose of short-term (200 days) display. The AAZPA opposed the approval of Toledo's required FWS import permit and joined the WWF in trying to stop the zoo from exhibiting the pandas on the grounds that the purpose of such exhibition was primarily commercial. The resulting lawsuit against the Department of the Interior—*World Wildlife Fund v. Donald Hodel* (1988)—was heard in the U.S. District Court for the District of Columbia.

The modern history of giant panda exhibitions dovetailed conservation, economic, and diplomatic interests. Thanks in part to the WWF's striking logo, the giant panda came to symbolize the plight of endangered species worldwide, and the public eagerly embraced panda-related products and exhibits. Capitalizing on the panda's endearing presence, Edwin Bergsmark, president of the Toledo Zoological Society, independently negotiated a loan of two pandas—Le Le and Nan Na—in early 1988 from China. He recognized that the pandas' visibility would draw national attention and additional income for facilities renovation at his institution.

For their part, the Chinese government exploited this western demand, providing occasional panda loans in exchange for economic support of Chinese conservation programs and political goodwill. The PRC had given President Richard Nixon the two giant pandas exhibited at the National Zoo. Similarly, the Bronx and San Diego zoos had been able to get panda loans from the PRC in exchange for financial support of a conservation research station in China.[57] In the case of Toledo, Wang Meng Hu, a member of the board of directors of the Chinese Wildlife Conservation Association and the person with greatest direct control over the panda reserves at the time, reminded Americans that China lent pandas "only for the friendly relations of the two countries and the two peoples" and warned that the opposition by the WWF, the AAZPA, and the NYZS would "affect the future cooperation between the two sides in their protection of the giant pandas."[58]

William Dennler, the Toledo Zoo's executive director, was surprised by the panda loan, which had been arranged largely by a board member of the zoo's nonprofit zoological society, and tried unsuccessfully to enlist the support of the AAZPA before seeking an import permit. In fall 1987 the AAZPA board had created a giant panda task force to address "substantial concern" about how the San Diego Zoo had used a pair of pandas on short-term loan. This task force marked the next step of an effort undertaken in 1985 to develop short-term guidelines for its members, a process begun at the urging of George Schaller, a biologist employed by the NYZS, whose experiences working on panda conservation in China led him to predict an increase in panda "rentals."[59] In February 1988 the task force met with representatives from the WWF, Department of the Interior, and the Species Survival Commission of the IUCN. The guidelines created at this meeting included returning pandas quickly to the wild, not exhibiting breeding-age pandas, and creating a coordinated breeding program. On March 5, 1988, the board of directors adopted as "mandatory standards" this position statement on giant pandas. In the same month Frank Dunkle, the director of FWS, wrote Wagner, asking for the AAZPA's "assistance in thwarting the growing number of Giant Panda short-term loan applications." Dunkle wanted more time to assess both the impact of these loans and the Chinese use of funds received from the loans.[60]

In response to the Toledo Zoo's request for support, the AAZPA instead referred to the new policy and expressed serious reservations about the zoo's planned exhibition.[61] The AAZPA wanted to know, for example, whether "this activity [would] truly benefit panda conservation and not encourage further exploitation"; how this "exchange [was] any more 'non-political' than any previous exchange"; and why the Toledo Zoo wanted to exhibit the pandas if they were "truly not making a profit on the panda exhibition." The AAZPA understood the zoo's general need to boost attendance and sales revenues but felt that "it should not be done at the expense of an endangered species" and that zoos should be "honest" and not "deny" their obvious desire to use pandas to increase attendance. The AAZPA was also disturbed by a Chinese decision to switch pandas at the last minute, because "the new animals proposed are clearly of breeding age supposedly . . . captured on December 17, 1983, and January 5, 1984, just before China ratified CITES on January 23, 1984." The convenience of these capture dates seemed "too much of a coincidence." Thus, the AAZPA rejected the zoo's request to support its permit application to import the pandas. Nonetheless, Toledo Zoo officials traveled to China to arrange the importation. On May 6, FWS issued the importation permit—one for the ESA and the other for CITES—even though the agency had not yet examined the Chinese export permit, as required by law. In a nod to the local political significance of the loan, it was rumored that an Ohio congressman had lobbied the FWS on his state's behalf.[62]

When the pandas arrived at the Chicago airport on May 14, 1988, the FWS noticed that the permit mistakenly referred to the pandas as "pre-Convention" animals; because the actual pandas were born after CITES, they technically required a different kind of permit. The FWS, however, allowed the Toledo Zoo ten days to produce a proper permit, which it did. Once on exhibit, the pandas drew an average of ten thousand visitors a day. In the same week that the pandas were due to arrive in the states, the WWF and the AAZPA filed a lawsuit against the FWS, first seeking an injunction to prevent the importation altogether on the grounds that they "failed to make the requisite findings" for issuing the permit. When this injunction request was denied, the AAZPA and WWF filed again, asking the court to order the FWS to stop the "further exhibition of the pandas," confiscate them, return them to the China, and "declare the permit invalid."[63] In the revised filing, the AAZPA stressed that the FWS had failed to consider the "commercial" use to which the pandas would be put.

In justifying to the court why it should exhibit panda bears, the Toledo Zoo rolled out a list of familiar educational benefits. It had created extensive educational material to accompany the visit and coordinated the exhibition with a cultural exhibit on China. Essentially, the zoo took a position often articulated by the AAZPA itself that firsthand exposure to endangered species boosted public support for conservation efforts. In exchange for permission from the Chinese to import the pandas and to comply with the in situ conservation programs pushed by the AAZPA, the "Zoo promised a minimum of $300,000 in equipment and vehicles as well as a promise of future joint activities with a panda reserve in China."[64]

Surprisingly, the Toledo Zoo made no secret of its plans to use the animals for economic gain. According to one informed observer, "the Toledo Chamber of Commerce expected a seventy-seven-million dollar bonus for the city." Dennler wrote a fund-raising letter to zoological society members, informing them that "after months of negotiation on the highest diplomatic levels" the zoo had "final approval *to bring a pair of critically endangered giant pandas from China for an historic visit to our own Toledo Zoo this summer!*" He enthused that this was "good news for Toledo" because the pandas would draw tourists from throughout the Midwest to the city. And it was "good for the Zoo too," because the pandas would be housed in a specially renovated elephant house that would later become "a fantastic new naturalistic facility" for the zoo's primates. An additional panda admission fee would raise some of the money for these improvements, but the zoo also needed help from members, so Dennler pitched discounted family membership plans and urged his readers to "come be part of 'pandamania' at the zoo."[65]

In defense of the FWS's decision to approve its permit, the Toledo Zoo leveled a number of stinging accusations against the WWF and the AAZPA. Noting that the WWF had made a substantial amount of money from some of the panda loans to zoos, as well as the image of the panda itself, the zoo accused the organization of trying to secure a monopoly on the panda through its relationship with China: the "giant panda *is not,* and should not become, a property which is subject to the exclusive control of the World Wildlife Fund."[66] Of interest is that it was precisely this financial relationship to the panda that the WWF claimed gave it standing. The WWF suffered an injury from the permit because "its members have a financial and philosophical stake in panda conservation."[67] Regardless, the Toledo Zoo questioned the WWF's conservation commitment by claiming that it opposed "China's carefully planned programs designed to enhance the propagation of the panda." Similarly, the zoo presented the AAZPA as a reckless organization violating its own procedural rules not for the purpose of conserving pandas, but to damage the Toledo Zoo. In a distortion of the record, the zoo claimed that the AAZPA had "hastily" drafted its panda policy. In fact, the *AAZPA Newsletter* had been reporting on the evolving policy for several years. The zoo's attorney also implied that the AAZPA position was elitist, noting that the decision on the panda question was made by "only [the] 15 person board of directors" rather than by the "thousands of members of the AAZPA or its 140 member zoos."[68]

Supporting the zoo in efforts to portray the WWF and AAZPA as riding roughshod over local interests, the state of Ohio intervened on the Toledo Zoo's behalf, arguing that Ohio had an "overriding and direct interest in the outcome" of the case because the "citizens of Ohio will be irreparably harmed" if the Toledo Zoo was prevented from exhibiting the pandas.[69] Thus, the zoo not only argued that the FWS had followed the permitting process, but also painted itself as an aggrieved, local public institution under attack from powerful outside organizations.

On the other side, the WWF and the AAZPA argued persuasively that the FWS failed to make sure that importing the pandas was legal under CITES. At issue here was the requirement of CITES Article III(3c) that "the specimen is not to be used primarily for commercial purposes." The court agreed that "it was undisputed that before the agency may issue a permit for the importation of an Appendix I endangered species, its Management Authority must examine the details of the proposed exhibition and make a finding that the animals are not being used for 'primarily commercial purposes.'" Unfortunately for the Toledo Zoo, it not only was charging an additional admission fee to see the pandas, but also had neglected to inform the FWS of this fee in its permit application. The court concluded that there was "no evidence in the administrative record that the agency either considered the commercial nature of the Zoo's exhibition, reached a conclusion that it was noncommercial in nature, or that the import was not 'primarily for commercial purposes.'"[70] On the basis of evidence that the Toledo Zoo was using the pandas for commercial gain, the court ruled in favor of the WWF and the AAZPA and ordered the zoo to stop collecting the additional panda-viewing fee.

Although the AAZPA won this case and sent a clear message about their continuing commitment to conservation, they did so at some peril to their organization. Reflecting on the lawsuit in 1989, President Palmer Krantz noted that a "considerable number" of AAZPA members had expressed concern about the lawsuit. If the AAZPA was going to sue such a high-profile member, who was next? William Reilly, the president of the WWF, however, reminded Wagner that, though their joint action may not have been well received by some zoos, the association had put itself on "record" by "taking a tough stand for conservation over financial interests." Defenders of Wildlife agreed, praising Wagner for his organization's stand and reminding him that a member of the board of directors of Defenders had served as the WWF's lead attorney. As Krantz himself said, after the Toledo lawsuit, "[N]o one can question the commitment of the AAZPA towards conservation."[71]

For his part, Wagner had eagerly embraced the WWF during to the panda case. Writing to his board members in 1987, he noted that "Santa Claus" had "arrived early" for him because the WWF had joined the AAZPA in the "related organization" category and had made the AAZPA a "professional associate" in its own organization. He noted that, regardless of his additional achievements over the past year, "the relationship with the WWF" was the one for which he was "most proud." Wagner looked forward to cooperating on a variety of conservation initiatives. Russell Train, chairman of the WWF board in 1987, believed that zoos were an ideal vehicle to get their wildlife conservation message to citizens and hoped to share education programs, cooperate in research, and work together on zoo and aquarium programs for the public.[72]

Behind the Toledo panda case was the resentment that a smaller zoo felt toward the long-standing power of larger zoos in the association. If New York could exhibit giant pandas, Toledo should be able to have them as well. To deny Toledo its right to get pandas while allowing New York to have them would forever keep the facilities unequal in prestige. Ohio politicians and zoological society members were well aware that an inferior zoo meant a less prestigious city. If Toledo wanted to compete with New York, it had to be aggressive.

Underlying the killer whale and panda cases was the desire of zoos to showcase attractive, crowd-pleasing animals to bring in needed revenue. Elaborate, spacious exhibits that best met the physical and emotional needs of animals required money. The financial rewards that accompanied such attractive animals made the risks of lawsuits from animal protection advocates worth the gamble. As an organization dedicated to protecting its members' interests, the AAZPA understood and sympathized with these financial considerations. The AAZPA could comfortably defend Sea World's taking of killer whales because there was no scientific evidence that the species would be threatened by the action. The Toledo Zoo, by contrast, acted contrary to an AAZPA policy designed to protect the giant panda. No animal protectionists objected to zoos' display of giant pandas, but the AAZPA could not allow the zoo community to ignore its conservation responsibilities. The AAZPA wanted to protect its members' financial interests, but not at the expense of the association's conservation image. The subtle distinctions that made Sea World's killer whale display "educational" and Toledo's giant panda exhibit "commercial" may, however, have been lost on the average citizen, just as the argument that euthanizing surplus animals was a humane way to preserve species became increasingly difficult for the HSUS to accept.

HSUS Begins to Change Heart

The HSUS began the decade self-consciously distinguishing itself from anti-zoo organizations such as United Action for Animals. It anticipated working cooperatively with the AAZPA against roadside zoos and lending its support to AAZPA programs. For the most part, HSUS officials, especially Pressman, played this role, but new HSUS personnel and changing circumstances during the 1980s increased tensions between the two organizations. In 1982, John Grandy, who had worked for Defenders, became the HSUS's vice president of wildlife and environment. In 1986, David Herbet replaced Pressman as the HSUS's captive wildlife specialist, and, though he promised to "continue to work together" with the AAZPA, his relationship with AAZPA directors was more conflicted and less personal than had been Pressman's.[73]

By the end of the decade, the HSUS's support for AAZPA zoos was less clear, largely because of the effect of SSPs. On the one hand, in 1989 the HSUS and AAZPA cooperated closely on a petition to Animal and Plant Health Inspection Service to strengthen several animal care standards under the AWA. In the same year, the HSUS also invited the AAZPA to work with it and other welfare groups on developing the educational program standards required by the MMPA amendments of 1988.

But Wagner was not completely comfortable with the HSUS. He had doubts about whether the HSUS fully understood "the role of modern zoological parks and aquariums," occasionally telling Herbet that he needed to "become more familiar with the AAZPA." Wagner chastised Grandy for the "gross inaccuracies" in his June 1988 lecture at a Zoocheck conference and expressed his surprise that he knew "so little about an organization whose efforts [the HSUS] purports to have been 'watchdogging' for many years." Nonetheless, at a 1989 meeting about the MMPA's educational programs, Grandy complimented the AAZPA's political activity, and Herbet said that he felt "more comfortable working with the AAZPA" than with the other groups present (Project Monitor, API, Center for Marine Conservation, and American Cetacean Society). The AAZPA's Kristin Vehrs thought that Grandy and Herbet "were being very supportive" of the AAZPA, but she was nonetheless "unsure" about "HSUS's overall intent."[74]

In 1989 Grandy published an article titled "Captive Breeding in Zoos: Destructive Programs in Need of Change," in which he expressed significant reservations about SSPs. The appearance of Grandy's anti-SSP article chilled the relationship between the HSUS and AAZPA. Chaffee regretted the "misconceptions" it contained but doubted that Grandy would publish an AAZPA response for HSUS readers. AAZPA leaders were clearly unsettled by the

HSUS's criticisms of SSPs. If the most moderate of animal welfare groups took this position, what could the AAZPA expect from the more radical protectionists who were already calling zoos "pitiful prisons"?[75]

The AAZPA could hardly turn its back on SSPs because they had become so central to the zoo community's mission and, indeed, to its relationship with the federal government. In 1988 the AAZPA's legislative committee chairman, Edward Schmitt, claimed that the ESA and MMPA, far from "restrict[ing] the activities of zoos," actually helped zoos breed and exhibit endangered animals. Schmitt stressed that SSPs in particular had been "encouraged and reinforced by" this legislation, which was the "zoos' major ally in developing their conservation role." Conway had anticipated some of the implications of this relationship when he "reflected on the nature and aspirations of zoos" in 1981. Conway proposed that the only way that zoos could maintain their long-term commitment to vanishing species, a commitment that he described as a "shimmering robe of light and goodness," was with governmental involvement. Antizoo critics often accused the AAZPA of representing the zoo "industry," but Conway doubted that the private sector could be counted upon to sustain zoos' conservation mission. He reminded Chaffee, then president of the association, that commercial zoos such as the Busch development in Houston had often "vanished upon a whim." He warned that zoos dependent upon the marketplace were "exceptionally insecure." The other needs of a corporation could easily come before those of the animals. Even the vaunted San Diego Zoo nurtured its "governmental connections" while other clearly commercial institutions like Sea World had developed programs that won federal support. In Conway's view, only with the help and recognition of government officials could zoos work toward their ultimate purpose of becoming "the leading centers of environmental education."[76] If zoos wanted to preserve species, they needed the moral weight of the government to assure the continuity of the project, because the marketplace was too likely to abandon the project when profits sagged.

The 1980s had been a decade in which the AAZPA gradually elevated its visibility and credibility—in the courts, with its SSPs and other conservation activities, and by accrediting more than 150 of its member institutions. In its actions, the AAZPA always claimed to be promoting the welfare of animals, individually and as a species. But it also sought to protect the public image of zoos in general, for example, by signaling its opposition to the Toledo Zoo's commercial exploitation of endangered animals. Wagner and the AAZPA leadership had taken a real gamble in the Toledo Zoo case. If the court had decided that the Toledo Zoo was *fundamentally* a commercial institution, the ruling would have both made the zoo ineligible to exhibit pandas and might have threatened all zoos' ability to import endangered species for any reason. In the years ahead, the AAZPA would return to this issue as animal protectionist organizations filed a number of lawsuits challenging zoos' noncommercial status.

By the end of the 1980s it was clear that the AAZPA was caught in a paradox. It had actively adopted species preservation programs to support its fundamental conservation mission, restock zoo animals, and appease critics in animal protection organizations. The WWF enthusiastically and formally advocated the AAZPA's conservation mission and its related programs. Defenders of Wildlife also publicized the SSPs and affirmed the "good work many zoos are now doing in the field of endangered species." The Audubon Society similarly applauded this effort in 1989, calling zoos "a lifeboat for wildlife."[77] Saving an entire species even within the zoo population, however, required that zoos sacrifice individual animals in various ways. They had to rid themselves of older animals to make room for new ones, as seen in the Detroit tiger case, or collect wild animals to increase the number in captivity, as was evident in Sea World's bid for the killer whales. Even though the HSUS initially supported their euthanasia decisions, new leaders expressed growing doubts about SSPs. Ironically, then, the AAZPA's most idealistic conservation initiative to date inspired renewed attacks by protectionists on their members.

FIVE

Animal Welfare Issues Revisited

Twenty-two years after the passage of the Animal Welfare Act (AWA), animal welfare organizations still felt frustrated by what they perceived as the continued and unpunished abuses by the animal exhibition industry. In response to complaints that the Animal and Plant Health Inspection Service (APHIS) was not adequately enforcing the AWA, a House subcommittee held fact-finding hearings in 1992. Dr. Elliot Katz, president of In Defense of Animals (IDA), testified about the sometimes callous ways that zoos treated animals. In particular he cited the case of Timmy, a lowland gorilla who had been isolated for many years at the Cleveland Zoo before developing a close and loving relationship with a female gorilla named Katie. Unfortunately, Katie proved to be sterile, and so, in the larger interest of preserving the species, the Cleveland Zoo, under prompting from the American Association of Zoological Parks and Aquariums (AAZPA), decided to send Timmy to the Bronx Zoo, where he could mate with several female gorillas. Despite recommendations by Diane Fosse and other primatologists against separating the gorillas, Timmy was moved to the Bronx Zoo, where he flourished. The separation ended badly for Katie, however, who was repeatedly attacked by her new companion, Oscar. When Oscar finally bit Katie's foot so badly that part of one toe had to be removed, Katz's group threatened to sue, and the zoo separated the gorillas.[1]

Katz found Katie's suffering particularly sad because it could have been prevented if a federal court had supported the IDA's 1991 lawsuit to halt Timmy's transfer, which they argued would "result in needless pain and risk" to an endangered species. The IDA sought to block the transfer on the grounds that Ohio's anticruelty transportation statutes should take precedence over a U.S. Fish and Wildlife Service (FWS) permit, because the Endangered Species Act (ESA) specifically excluded animals such as Timmy, who had been born prior to December 28, 1973, *and* was not being held for "commercial" purposes. Disregarding both Congress's and the FWS's clear distinction between public zoos and commercial enterprises, the court held that federal law (in this case the ESA) preempted state law because Timmy was "held by a zoo, which would qualify as a commercial activity."[2] This may have been the only time since 1973 that a zoo was relieved to have the government identify it as a commercial institution.

Timmy and Katie's saga embodies several political themes that preoccupied zoos and animal protectionists during the 1990s. As the federal court's decision indicated, the question of what kind of animal exhibit constituted a "commercial" activity under the ESA and related regulations remained somewhat open to interpretation. In the Timmy case, a zoo benefited from the court's interpretation, but in most situations such an interpretation would have been disastrous. Because for-profit zoos were becoming a larger element within the zoo community, the AAZPA was particularly keen to convince the government to give them the same "noncommercial" status enjoyed by public zoos. The AAZPA also began to feel further negative side effects from successful Species Survival Plans (SSPs). First, because the efficient propagation of captive animals required shifting them from zoo to zoo, the AAZPA could expect an increase in cases like Timmy's. The scientific management of captive populations did not always sit well with activists, who increasingly sought to protect animals as individuals, rather than as representatives of a species. As a result, zoos continued to come under attack, despite a record of actually supporting tougher animal welfare rules. Ironically, when protectionists themselves gained control of captive wild animals, the results could be disastrous, as we will see.

Maintaining the image of zoos as humane, conservation institutions in the face of these themes—the inclusion of for-profit members, the continued pressure of animal welfare organizations, and the difficult problem of surplus animals—required the AAZPA to make several internal changes. In 1990, the association's newsletter was renamed the *Communiqué*. In 1992, the board selected a new executive director, Sydney Butler, who gave the association a bolder and more marketing-oriented public face. Butler brought nonzoo environmentalist credentials to the association, having served as the vice president for conservation at the Wilderness Society (1987–92), and AAZPA president Steve Taylor welcomed Butler, believing that "under Syd's leadership, the role of zoos and aquariums in conservation will be much better recognized by our government leaders."[3] Two years later (1994) the association renamed itself the American Zoo and Aquarium Association and adopted the shorter acronym AZA. By 1996, its newsletter was transformed into a glossy magazine with articles intended as much for a general audience as for members. The AZA was maturing, growing more confident, and developing a clear identity as an organization dedicated to the public interest. Its political struggles, however, were far from over.

For-Profit Zoos and the Question of Commercial Status

From its incorporation as an independent interest group in 1972, the AAZPA's relationship with its for-profit members had been conflicted. The animal welfare movement's attacks on small, for-profit roadside zoos made it politically expedient for the organization to disassociate itself from such institutions, and by the 1990s a robust accreditation program effectively excluded these zoos. More problematic, however, was language in the ESA, the Marine Mammal Protection Act (MMPA), and Convention on International Trade in Endangered Species of Wild Fauna and Flora (CITES), which prohibited permits for the "commercial" display of endangered animals. Because the AAZPA desired to include progressive, for-profit zoos among its members, it did not want government authorities to interpret "commercial" so broadly that it included all for-profit zoos. Rather, the AAZPA sought, and essentially gained, an interpretation that considered whether the animal's display would contribute to captive propagation, public education, or scientific research. Nonetheless, as the Toledo Zoo panda case demonstrated, the AAZPA took quite seriously these laws' underlying philosophical opposition to the exploitation of endangered animals, and it proved willing to oppose its own members who acted contrary to the interests of protected animals. Many animal welfare groups, not surprisingly, wanted the government to interpret "commercial" to include all for-profit zoos. At the margins, these groups perceived *any* display of animals as exploitative and therefore commercial, and they sought to entangle all zoos in the commercial category by arguing that the transportation of animals was a commercial activity. For its part, the AAZPA tried to both protect its for-profit members and prevent "bad" zoos from enjoying those protections.

Animal rights groups have been frustrated in their attempts to use the courts to force the FWS to use the commercial exclusion language to prevent zoos from moving endangered animals as they desire, partly because the courts seem to interpret that language in a manner that benefits the zoo in question. In the case of the Cleveland Zoo, a court found the zoo protected by its commercial status, and in a case a few years later a different court found the Milwaukee Zoo protected by the noncommercial nature of its animal transfer. In 1992, the Humane Society of the United States (HSUS) sued the Department of the Interior and the FWS because they allowed the Milwaukee Zoo to transfer Lota, an endangered Asian elephant, to the Hawthorn Corporation, which planned to use her in a circus. After thirty-six years in captivity, Lota had begun to "exhibit aggressive behavior toward the zoo's other Asian elephants," particularly an older elephant named Tammy, whose health and safety were "jeopardized by Lota's actions." In November 1990, after behavior modification efforts failed, the zoo decided to giver her to the Hawthorn Corporation. In light of the endangered status of Asian elephants, Lota was potentially subject to regulation under both CITES and the ESA. The FWS, however, in 1991 issued a CITES certificate designating Lota as a "'pre-Convention' animal exempt from CITES' import and export restrictions," which included "interstate or foreign commerce in the course of a commercial activity." The HSUS wanted to force the Secretary of the Interior to require the Hawthorn Corporation to apply for a conservation permit, because Lota would likely be performing in a circus. The HSUS suspected that Hawthorn could not meet the conservation requirements and that Lota would be used "contrary to efforts 'to enhance the propagation or survival' of Asian elephants," a charge that the corporation took as a broader attack on their ability "to give elephant rides and circus performances." If one considered Lota's ultimate destination, the HSUS seemed to have a strong case. In fact, in agreeing to hear the case, the court disagreed with the defendant's argument that only purchases and sales constituted "commercial activity." The AAZPA feared that the case "could change the meaning" of this very important phrase. Fortunately for the zoo community, the court ultimately focused on the elephant's *exchange,* rather than on Lota's eventual fate.[4]

The court's decision for government, upheld on appeal in 1995, supported the FWS's interpretation of commercial activity. It noted that specific provisions of the ESA's "captive-held exemption" dictated that endangered species held "in a captive or controlled environment" prior to December 28, 1973, were exempt from the prohibition on export or import of endangered species "*provided,* that such holding and any subsequent holding or use of the fish or wildlife was not in the course of commercial activity." Obviously, the transfer of Lota to a company that provided circus animals seemed to violate the latter portion of this provision, but the court chose to ignore the "subsequent" use of Lota and focus instead on the nature of the transfer itself. Because the zoo was giving the elephant away for free, the court interpreted the transfer as a noncommercial transaction. In reaching this conclusion, the court cited an ESA description of "commercial activity" as "all activities of industry and trade, including, but not limited to, the buying or selling of commodities." Historically, the FWS had interpreted "industry and trade" to mean only animal transactions in which the seller made a profit, and in many situations zoos were simply loaning animals to each other, in which case no profit was realized. In light of the FWS's historical interpretation of "commercial," as well as the legislative history of the ESA in which Congress had never seen the need to revise the somewhat "ambiguous" definition of "commercial activity," the court concluded that the transportation of Lota "does not constitute 'commercial activity.'"[5]

Together, the cases of Timmy and Lota illustrate how zoos have been able to defend their acquisition and transfer policies under cover of being noncommercial institutions with special legislative and treaty status relating to trade in endangered species. Courts have generally avoided the larger question of whether zoos *are* commercial, focusing instead on narrower debates about whether specific activities (e.g., charging an extra fee, transferring an animal) are commercial in nature. Fortunately for zoos, the courts have basically followed Congress's and the bureaucracies' lead in viewing public zoos as noncommercial institutions. The AAZPA, in fact, worked hard to ensure that legislators and bureaucrats explicitly distinguished between commercial and zoological institutions, recommending, for example, that in a 1973 rule pertaining to the importation of birds language be added to distinguish "zoological birds" (those "imported by a zoological garden, which qualifies as a non-profit tax exempt organization") from "commercial birds."[6] Animal protection groups understandably felt frustrated by this status, because the typical zoo charged admission, sold souvenirs, and generally engaged in a range of apparently commercial behaviors. On the other hand, the AAZPA could reasonably argue that it proactively defended animals against government decisions that might benefit zoos. This kind of action had been evident in the Toledo panda case, and it became even clearer in the AAZPA's subsequent work with the FWS to develop a panda policy.

The AAZPA and the FWS spent a decade working out a policy for the importation and breeding of giant pandas. The evolution of this policy highlighted the AAZPA's desire to assert control over its members by making participation in an SSP a permit requirement. At the same time, in its conversations with the FWS, the AAZPA also sought to protect its for-profit members from potentially damaging details in the proposed policy. The result was a policy that gave for-profit zoos the ability to import pandas, as long as they participated in a range of more sophisticated conservation activities favored by the AAZPA.

Like the AAZPA, the FWS became concerned by the late 1980s that in the absence of a clear panda importation policy, numerous American zoos (as many as thirty) were seeking to acquire short-term loans of pandas from China. The AAZPA and the FWS agreed that, because no panda SSP existed, such loans, while technically permissible, threatened the wild and captive populations of pandas by doing nothing to increase their propagation. Moreover, the CITES secretariat in Lausanne, Switzerland, had reacted negatively to the FWS's approval of the Toledo Zoo's 1988 panda permit and requested that the service "reconsider its permitting action." As a result of these concerns and the court's decision in the Toledo case, the FWS suspended "all future" panda imports and began the process of revising its panda exhibition and loan policies. Simultaneously, the AAZPA began to develop a giant panda conservation action plan that would bring twenty-nine zoos together in a cooperative breeding program.[7] Once again, it is clear that the AAZPA wanted to control its members' activities, even as it desired to make an endangered species available to them; at the same time, the Chinese government wanted to loan its pandas, and the FWS was under pressure to take action, so the AAZPA stood a good chance of getting what it wanted.

In a savvy move, the AAZPA negotiated directly with China, thereby setting the parameters for the FWS's eventual panda policy. It had laid the groundwork for such negotiations when Robert Wagner attended the First International Conference on Wildlife Conservation in China, held in Beijing in 1987, and returned hopeful that meetings with a newly formed Chinese Association of Zoological Gardens would lead to "positive cooperation" on breeding programs. Seven years later, on January 14, 1994, the AZA signed an agreement "with Chinese officials" stating that the "Chinese [would] support a long-term giant panda captive-breeding cooperative plan in North America as presented by [the] AZA." The AZA's plan called for a giant panda SSP, propagation research, and American financial support for a range of Chinese activities, including professional training and habitat conservation.[8] By securing such a detailed agreement with the Chinese government, the AZA essentially assured that the FWS would adopt a very similar policy.

The AZA also worked to ensure that the FWS's policy accommodated for-profit zoos. The most controversial elements of the government's evolving panda policy were the related questions of whether for-profit zoos could receive loan permits and what would happen with zoos' increased revenues because of the loans. For-profit institutions hoped that they too would be allowed to exhibit pandas, arguing that as AZA members who met the association's accreditation standards, they should be exempt from their designation as for-profit institutions and treated the "same way as not-for profits." The for-profit zoos were also unhappy with a FWS proposal that net profits from panda exhibits (e.g., from additional ticket sales, souvenirs, and so forth) go toward panda conservation, because calculating such profits under the FWS plan would require the for-profits to provide a public accounting of their revenues.[9] In something of a reversal of its position in the Toledo Zoo case, the AZA supported its for-profit members.[10] When the FWS published its final panda policy in 1998, the AZA had once again succeeded in protecting its members. The FWS ruled that "any public, private, non-profit, or commercial (profit-making) institutions, organizations, and agencies will receive consideration" for a permit but noted that for-profit institutions would have a harder time securing a permit because it would be "more difficult for the Service to find that the specimen proposed for import is not to be used primarily for commercial purposes," because for-profit institutions' ultimate responsibility is to their shareholders rather than to the public.[11] On the issue of determining a zoo's net profit from panda exhibition, the FWS compromised, allowing zoos to hire an accountant to determine the amount, thereby keeping for-profit zoos' financial records private.

Whether zoos are "commercial" institutions is ultimately a political question, one likely to be debated for as long as zoos wish to import or exchange endangered animals. The AZA recognizes that, under the CITES convention, the distinction between commercial and noncommercial operations is "critically important" to zoos, and it has long argued that zoos are noncommercial, even when not public; however, just as the FWS's definition of "conservation activity" has evolved, so too the definition of "commercial" activity is subject to reinterpretation. In 2002 the CITES secretariat attempted to define zoos and aquariums as commercial operations, and again at the 2004 CITES Conference of Parties a participating nation proposed a resolution to require that import permits for Appendix I animals (the most endangered) be granted only when the transfer of the animals did not involve a commercial transaction. In other words, if a party profited somewhere between the capture and final delivery of an endangered animal, an import permit could be denied. This resolution would have superseded an existing one that prohibited imports only when an animal's "final use" would be commercial. If adopted, the resolution would have effectively halted zoos' importation of wild-caught endangered animals. The AZA defeated the 2004 proposal with an alternative that directed CITES to "conduct a review of trade" practices and their conformity to the CITES convention.[12] In view of the AZA's past success in this area, it may feel confident about the future.

Welfare Issues

Underlying the challenges to Lota's and Timmy's transfers were concerns about the physical and mental well-being of the animals. In these and numerous other cases, animal welfare organizations argued that zoos were contributing to the "abuse, neglect, suffering, and death" of animals. By 1992 the HSUS had become so frustrated by the frequency and severity of these problems that it reevaluated its standing policy to "work with" those zoos committed to endangered species breeding and meaningful public education. Concerned about what it perceived as a gap between the "rhetoric and reality" of American zoos, the HSUS declared that many zoos needed to close permanently. It also reaffirmed its desire to continue cooperating with the AAZPA, but it strongly cautioned that the "good" zoos needed to break the code of silence and publicly condemn crimes committed by their brethren.

The HSUS was not alone in calling for the end of some zoos. Public figures as disparate as Jacques Cousteau and Washington's secretary of state, Ralph Munro, declared that zoos should not keep marine mammals in captivity at all. Increasingly bold animal rights organizations went even further. The Animal Rights Connection (California), for example, posted fliers with the headline "Imprisoned Without Trial—Abolish Zoos." In smaller print, the flier "debunked" zoos' conservation message by claiming that, far from saving endangered species, "zoos are a major cause of extinction"; that contrary to the touted success of reintroductions, they have been "dismal failures"; and that zoos serve only to "relieve" the public of guilt about extinction. Just as in the mid-1970s, some within the AAZPA began to use a rhetoric of uncertainty and doom to describe the increasing "pressure from Animal Rights Organizations."[13]

Paradoxically, as the animal rights groups gained numerical strength, they lost political power, at least in their conflicts with zoos. The 1990s saw these groups pursuing zoos through the courts primarily for publicity's sake, but sometimes in a somewhat disorganized manner that resulted in decisions making further legal actions very difficult. The more radical groups also took dramatic but legally questionable actions against the captive animal industry, secretly videotaping abuses, breaking into research laboratories, and even destroying property. These actions helped fuel a backlash from a new Congress, and zoos quietly took advantage of legislation that provided the animal industry with legal protections from animal rights groups. It would be wrong, however, to conclude that the AAZPA ignored animal welfare considerations during this period. On the contrary, it actually pushed for stricter government enforcement of animal welfare laws.

Before an American citizen can take legal action against a public or private institution that injures him or her, the courts must grant that citizen "standing" to do so. As the Detroit Zoo tiger case demonstrated, animal protectionists have not always attained this standing. A variety of laws protect animals from a range of specified abuses and actions, but the humans who seek to use the court system to invoke those laws must show "injury" to *themselves,* not to the animals. Animal protectionists have encountered great difficulty gaining standing, particularly in suits against the biomedical industry.[14] Even when the courts have granted standing, they have generally rejected the plaintiffs' central complaints. For example, when IDA sought a restraining order to stop the movement of three gorillas from the Los Angeles Zoo in 1992, the court recognized the organization's standing but dismissed their case on the technical grounds that they had failed to file a sixty-day notice to sue.[15] More typically, the courts have found the animal protectionists' claims of injury unconvincing, reflecting the protectionists' haphazard legal strategy.

As many examples have demonstrated, animal protectionists often objected to zoos' decisions to transfer or dispose of their animals. In 1993 the organization Citizens to End Animal Suffering and Exploitation (CEASE) filed suit against the New England Aquarium and several government agencies to protest the aquarium's transfer of two dolphins (Kama and Rainbow) to a U.S. Navy research project. The lawsuit focused on a technical question relating to the MMPA permitting process, but the underlying concern was for the dolphins' welfare. Born at Sea World in 1981, Kama had been transferred to the New England Aquarium in 1986 for breeding and display purposes, but he did not get along with the other dolphins and was rarely displayed.[16]

CEASE attempted to gain standing in the case first by elevating Kama to the status of plaintiff and then by claiming injury in the form of its members' inability "to observe and study Kama" after his transfer. The first strategy found little favor with the U.S. District Court for Massachusetts, which pointed out that the "MMPA expressly authorizes suits brought by persons, not animals." The court was more receptive to the possibility that CEASE had standing, if at least one member could show that he or she had been harmed by the actions of the New England Aquarium. But what kind of injury did animal lovers face when a dolphin was removed from an exhibit? One could imagine that the activists faced emotional trauma at the plight of the animals, but CEASE initially borrowed a legal strategy from environmentalists and made the somewhat strange case that they suffered "aesthetic harm," which they argued was an "injury in fact" that was "concrete and particularized" as required by the Supreme Court.[17]

Although open to the possibility that CEASE had standing, the court complained that the organization made its injury arguments in a less-than-thorough fashion. First, CEASE argued that its members "suffered injury to their aesthetic" interests. Two of the plaintiffs stated that, during the time they were at the aquarium, they "attended dolphin shows and saw dolphins on public display there several times." The court, however, found the injury here vague, noting that the affiants had not "stated whether they have returned to the Aquarium" and "suffered injury because of Kama's absence" or even if they planned to return again. Moreover, because neither affiant was sure that he or she had "actually observed Kama" and indeed had taken three years to notice his absence, the court concluded that removing Kama would not cause the plaintiffs any aesthetic harm because there were other dolphins at the aquarium the activists could observe if they so desired.[18] The activists thus failed to gain standing not because the court ruled out the possibility of such standing, but because they failed to show how they were injured by Kama's removal.

Strangely, the HSUS repeated some of CEASE's mistakes in its attempt to prevent Lota's transfer to a circus company. Once again, a court treated a plaintiff's claim of "harm" with skepticism, finding little evidence to support it. In this case, one of the plaintiffs, a Milwaukee resident named Kay Mannes, claimed that "she had visited the Milwaukee Zoo several times between 1968 and 1992," where she had seen Lota and "found the experience 'both educational and aesthetically satisfying.'" By viewing Lota, she had "learned more about the Asian elephant" and "gained appreciation of endangered animals." As a result, she claimed to be harmed by the loss of Lota at the zoo because she had "lost the opportunity to learn about an endangered animal, the Asian Elephant." The U.S. Court of Appeals for the District of Columbia noted that, though it was true that "aesthetic interests such as the observation" of endangered animals might sometimes confer standing, Mannes had failed to explain why Lota *in particular* taught her about Asian elephants. The court noted that Mannes did "not claim that Lota was the *only* [original emphasis] Asian elephant at the zoo" because there were three other elephants on exhibit. Although the court could "imagine a situation where a frequent zoo visitor's systematic observation of an animal" might be harmed by its removal, that did "not appear to be the case here—or at least it has not been well pleaded." Like the plaintiffs for Kama, moreover, Mannes failed to indicate whether she intended to "return to the zoo to observe the elephants." Thus, the court concluded that her claimed injury did not constitute an "actual or imminent injury" as required.[19]

The court also affirmed that any injury must be individual, not collective. The HSUS argued, unsuccessfully, that their members were emotionally harmed by Lota's move. The society averred that many of its members "visited the zoo elephants before and after Lota's removal and were 'distressed' by her absence." They claimed that their members had been "emotionally injured," suffering such ill effects as "sleeplessness, depression, and anger" after Lota was transferred to Hawthorn. The court, however, followed *Animal Lovers Volunteer Association v. Weinberger* (1985), in which the court held that animal rights activists' opinions on animal harm could not stand for everyone's opinions on harming animals.[20] The court also noted that eighty of the eighty-three letters that the HSUS included in their affidavit claiming emotional injury came from society members who failed to mention whether they had ever visited the zoo, and "none of these make any reference to viewing Lota there." The court remained unconvinced that an individual had suffered any kind of injury as the result of Lota's transfer.

Setting aside the issue of standing, these cases revealed the extent to which zoos and animal protectionists fundamentally disagreed about what constituted concern for an animal's welfare. From the zoos' perspective, Lota represented a threat to the welfare of other elephants, Kama's antisocial behavior condemned him to a life in captive isolation, the Detroit tigers were suffering from age-related ailments, and Timmy was unable to fulfill his natural urge to mate. Nevertheless, animal protectionists saw each situation quite differently, characterizing the zoos' actions as callous and cruel. None of the cases, however, involved the kinds of obvious abuses—mistreatment, confinement to tiny cages, or vicious exploitation—that had prompted Congress to amend the AWA in 1970. If animal protectionists were willing to characterize seemingly humane treatment of animals as "abuse," zoos would be subject to almost constant harassment. In fact, the tactics of more radical organizations in the movement had begun to attract the attention of legislators, and zoos soon would enjoy stronger legal protection from these groups.

Because the AWA primarily affects commercial agriculture interests, all related legislation is referred to the House and Senate agriculture committees and their subcommittees. These committees tend to be dominated by representatives from farm states, whose primary concern is protecting their constituents' ability to profit from animals. Pending legislation, even if its focus is on laboratory or performing animals, is examined closely for its potential impact on agribusiness or the "food and fiber" animal industry.[21] So, although zoos are somewhat of an afterthought for most committee members, they benefit from the committees' hostility toward the animal rights agenda. This hostility came to the surface in two congressional actions taken during the period when animal protectionists were pursuing zoos in the courts.

By the late 1980s, biomedical and commercial agricultural interests teamed up to seek legislative shields from radical animal rights activists. In response, the conservative Charles Stenholm (D-TX) and several cosponsors from the agriculture committee proposed the Farm Animal Research Facilities Protection Act of 1989 (HR 3270). The subsequent House Subcommittee on Department Operations, Research, and Foreign Agriculture hearing on Stenholm's bill quickly established that Congress had come a long way from the days when it lent a sympathetic ear to tales of animal abuse. George Brown, Jr. (D-CA), opened the hearing by characterizing animal rights activists as "terrorists," citing a 1989 U.S. Department of Justice report on terrorism that gave the Animal Liberation Front (ALF) responsibility for more than a hundred criminal acts. Brown was followed by deans of medical colleges, presidents of state farm bureaus, and representatives from agricultural animal interest groups, many of whom testified to the fear that ALF had created in the biomedical research community. At times, witnesses implied that animal rights activists were willing to harm animals in their cause, as in the case of ALF burning down a University of California—Davis veterinary diagnostic center.[22] To discourage such terrorist acts, Stenholm proposed increased criminal penalties for the offenders.

Seeking to avoid public identification with the animal research community, the AAZPA did not testify at the hearing, but it worried that the bill failed to include zoos among the facilities protected. Internally, AAZPA members had been discussing the fear that they too would be targeted by animal rights groups trying to stop research on animals. The research activities of zoos had come under occasional scrutiny by animal welfare groups in the 1970s, when comparatively few institutions had robust research agendas, but by 1990 the AAZPA had made "research" a central part of zoos' stated mission. For a time zoos hoped that they could stave off newer, more radical national animal rights based groups such as ALF with assurances that their research on zoo animals was undertaken only to benefit animals and was thus ethically justified. Animal rights groups, however, accept the theoretical arguments of Peter Singer and Tom Regan, among others, which posit that captivity itself is unethical. Both theorists and activists believe that animals should have rights that are similar to those held by humans at least in the core senses they have a right to live without excessive pain or death inflicted by humans and they should be free. Animals need more than just our pity and small amounts of welfare; they need rights afforded by the legal system and, if that fails, rescue. They are particularly concerned about any violence toward an animal and tend to focus on marine mammals and primates because they feel that their intelligence makes it particularly unfair to place them in tanks or cages.[23]

At least one zoo did come under attack by ALF—a group also motivated by the philosophical arguments of animal rights theorists. In 1988, keepers for Dunda, an elephant at the San Diego Zoo, allegedly hit her over the head with an axe handle repeatedly over the course of two days in an attempt to discipline her because she had been recently brought to the zoo and was not acclimating well. In response, the HSUS issued a statement condemning the beating as animal abuse and called for investigations. Meanwhile, ALF members "vandalized the cars and the homes of three zoo keepers who participated in chaining and beating" the animal, arguing that it took matters into its own hands because "no charges of animal cruelty" had been pressed against the keepers responsible for the beatings.[24]

Local animal rights groups also engaged in direct action protests and physical attacks on aquariums and dolphin facilities in the 1990s. For example, the Ocean Reef resort in Florida owned two dolphins—Bogie and Bacall (both female)—that it transferred to a temporary holding facility on the Indian River Lagoon, where unidentified activists illegally cut a net keeping the dolphins in the lagoon. Also in Florida, Richard O'Barry, who had trained six "Flippers" but became a dolphin-rights activist on Earth Day in 1970, was actively protesting the National Aquarium's dolphin-holding facility at the Hawk's Cay resort in Duck Key, Florida, in the winter of 1990. Baltimore's

National Aquarium had four dolphins that needed a home while their facility —in which two dolphins and a beluga whale died in 1981—was renovated. Officials sent the dolphins to Kevin Walsh, who had designed and directed the facility at the resort that faces the Atlantic Ocean. The dolphin holding tank, located in a canal, was about two football fields long. A net ran along the ocean side of the canal, preventing the dolphins from escaping, and the pens for separating and containing the animals were near the land. O'Barry had already protested against the dolphin facility at the National Aquarium. His planned protest on the Hawk's Cay facility was directed more at the National Aquarium's exhibition of dolphins than at Hawk's Cay as a facility. One day, O'Barry and fellow animal rights protesters arrived at the facility to protest the dolphin containment. They had alerted the media, who in turn, informed the local police, who told Walsh about the upcoming confrontation. Because Walsh appreciated the idea of protests generally and the importance of freedom of speech, he allowed the activists to enter the facility and stage their demonstration. The lack of response from either the resort guests or Walsh prompted O'Barry to leap into the water with the dolphins for at least some publicity. Unfortunately, among the dolphins were lactating females with calves who did not take kindly to strangers, and worse, a dominant male who headed straight for O'Barry in the pool, prompting him to quickly jump out of the water.[25]

Although O'Barry's activism was nonviolent, other unidentified dolphin activists behaved in more threatening ways. Activists defaced billboard's advertising Hawk's Cay by labeling them "dolphin killers." The owners of the facility were concerned enough about the graffiti that they hired off-duty Florida Marine Patrol officers to watch the dolphins and the hotel. When they were unavailable, they used an employee of the hotel as a night watchman. One night, dolphin rights activists snuck into the hotel lobby and covered the oriental rugs and statues with the heads, blood, and innards of fish. On another night, activists cut the net of the dolphins in the hopes that the animals would swim away. The dolphins, however, simply swam around nervously, afraid of the larger fish that had entered their enclosure.[26]

In a later instance, Richard Strahan, a regular protester at the New England Aquarium in the early 1990s, threatened to sink a New England Aquarium whale-watching ship by putting a hole in its hull. Strahan typically protested in front of the aquarium, urging visitors not to go on the trips because he felt that the boats harmed the whales by encroaching upon their space in the ocean. Frustrated by the lack of impact his protests were having, he eventually stood on the gangplank of the *Voyager II* and informed the ship's mate that he was "assessing the enemy," that it was "illegal to harass whales," and that he was looking for a place to "put a hole in the boat."[27]

At the 1990 AAZPA annual meeting in Indianapolis, Indiana, zoo and aquarium directors voiced their increasing concerns about the more aggressive attacks of animal protection groups. William Braker asked the "spies in the audience to get up and tell your side of the issue," whereupon Sue Pressman, who now worked for the Performing Animal Welfare Society, indicated that she still thought animal protection groups and zoos could work together, though her organization was particularly opposed to "animal performances." The year before, however, the Animal Protection Institute and seven other animal protection groups had sued to prevent the Shedd Aquarium, Braker's institution, from importing false killer whales from Japan. Well aware of activists' renewed attention to the display of marine mammals, Nicholas Brown, director of the National Aquarium, whose dolphins were the focus of much attention, warned that "today it may be dolphins; but tomorrow, it's everything else at zoos."[28]

Sensing the opportunity to seek federal protection from activists' physical attacks, Wagner urged the AAZPA to support Stenholm's bill. By March 1990, five bills aimed at imposing criminal penalties on animal rights activists trespassing in animal facilities were making their way through Congress. Braker advised all Marine Mammal Coalition members and the AAZPA to support a version of these bills that included zoos and aquariums to prevent another attack like the one at Hawk's Cay. At the AAZPA's "request," the "committee added language to include coverage of zoos and aquariums." When the House Judiciary Committee appeared ready to drop zoos and aquariums from the bill, the AAZPA and many member zoos wrote all committee members, urging them to retain the language.[29] Their lobbying was successful and the bill—eventually called the Animal Enterprise Protection Act of 1992—was passed by both houses and signed by President Bush. The bipartisan desire to protect the rights of medical researchers and companies to use animals in research worked to the advantage of American zoos.

The more moderate animal protectionists did not give up on legislative solutions, and in the same year that Stenholm's bill became law, they convinced the House Subcommittee on Department Operations, Research, and Foreign Agriculture of the Committee of Agriculture to hold a fact-finding hearing to determine whether animals in the entertainment industry were being adequately protected by the AWA. A report by the U.S. Department of Agriculture (USDA) inspector general supported the claim that the APHIS did not inspect facilities with sufficient frequency to ensure the humane care of animals.[30]

To dramatize the results of lax inspections, a parade of witnesses testified to the abuses of the animal exhibition industry. Actress Kim Basinger made a videotaped presentation in which she recounted the animal neglect she witnessed at roadside zoos, including "an elephant that was severely dehydrated and malnourished" and a "lion that was covered with some kind of skin disease." Katz recounted the sad tale of the Cleveland Zoo's Katie and Timmy, as well as that of Hannibal, a Los Angeles Zoo elephant that died of heart failure during an attempt to ship him to another zoo. An outraged zoo employee contacted IDA with the claim that "the attitude around the zoo was that Hannibal was going out dead or alive." Roger Caras, president of the American Society for the Prevention of Cruelty to Animals, claimed that APHIS was "totally inept and ineffective" in enforcing the AWA—a "joke, in fact." Caras called for a "totally new bureau dedicated solely to enforcing standards for the care of animals used for exhibition and other purposes" because "APHIS enforcement efforts have been so abysmal over such a long period of time that it is clear that drastic change is needed."[31]

In contrast to the pleas by animal rights activists for more regulation of zoos, the AAZPA opined that APHIS inspections were adequate and, in the case of its accredited zoos, practically unnecessary, because AAZPA standards exceed legal minimums. Butler, the AAZPA's new executive director, reviewed the accomplishments of zoos in the field of conservation and species preservation and argued that no further legislation was needed because the AWA already required that APHIS inspect zoos twice annually. From the AAZPA's perspective, APHIS simply needed more resources for enforcement and refinement of the existing regulations.[32]

Animal welfare and rights activists and zoos carefully laid out their arguments, but they did so for parties that were not especially interested in the concerns of either side. The real issue for several of the subcommittee members was not the problem of roadside zoos that Basinger recounted, the abuse of animals in zoos that Katz documented, or even the paltry funding of APHIS that the AAZPA claimed. The real concern for several members was the potential government regulation of rodeos. Mike Kopetski (D-OR) urged in his opening statement that the committee distinguish between those members of the animal entertainment industry that did a good job of maintaining animal welfare and those that did not. Establishing a theme for the hearing, he cited the Professional Rodeo Cowboys Association as an example of a group that took good care of animals and thus should not be seen as a problem. Wayne Allard (R-CO) stated that "he would not support any legislation" that shut down circuses or rodeos because there were "too many people in his district who rely upon the rodeo/exhibition industry for their livelihood." Similarly, Robert Smith (R-OR) reminded the subcommittee that since 1950 he had "actively operated a cattle ranch" and had been "involved in rodeos all of [his] lifetime." Thus, he concluded, "if this legislation [was] going to include rodeos it ought to be defeated." In the same vein, Pat Roberts (R-KS) pointed out that "the use of animals for various purposes is very, very prevalent on the high plains," and he spoke with pride about "the rodeos that have long been a key part of the rural fabric of America." Alone among the committee members, Roberts also put in a good word for zoos, noting their educational significance for his constituents and quoting approvingly from a local zoo director's letter about zoos and conservation work.[33] Few of the subcommittee members saw any need for fitting APHIS with additional legislative teeth, regardless of starving elephants and sick lions. The most important issue for these legislators was that no one should touch their constituents' horses and cattle. To committee members from western states, creating additional legislation looked like a slippery slope to interfering with rodeos.

In situations like the two above, the preoccupation of legislators with laboratory and farm animals protected zoos from the reach of their foes. At the same time, however, this preoccupation hurt zoos themselves. When Congress proved stingy with APHIS's enforcement budget, abuses at roadside zoos could go unchecked. When animal rights groups publicized such abuses, the AZA once again had to distinguish "good" zoos from "bad" zoos. This constant need to defend itself might have been avoided if non–AZA-accredited zoos were banned or if APHIS had a larger enforcement budget, but a Congress suspicious of the animal rights movement made both possibilities unlikely. Similarly, the USDA did not always promulgate rules that would adequately protect captive animals from inhumane conditions. A federal court case in the late 1990s revealed that, in such situations, the AZA would quietly side with animal protectionists, recognizing that its members were hurt when lax government regulations allowed shoddy zoos to abuse animals.

In 1996 the Animal Legal Defense Fund (ALDF) filed a lawsuit against the Secretary of Agriculture, seeking to compel the USDA to enforce the 1985 amendments to the AWA for primates.[34] In *Animal Legal Defense Fund v. Secretary of Agriculture* the plaintiffs charged that "the Agency's failure to issue a regulation promoting the social grouping of nonhuman primates is arbitrary, capricious, and an abuse of discretion in violation of the APA." Instead of issuing such a regulation, the USDA had allowed the facilities that housed primates to develop their own standards of care. To demonstrate the harm caused by the agency's action, the ALDF focused on the treatment of primates at the Long Island Game Park and Zoo, which housed its monkeys in isolation from one another in cement cages with bars, a situation similar to the kind of housing provided by many research facilities to nonhuman primates. The ALDF gained standing through Marc Jurnove, a longtime animal welfare advocate whom the U.S. District Court for the District of Columbia recognized as having a clear interest in the welfare of the game park's monkeys because he frequently visited zoos for "recreational purposes" and would therefore suffer if those animals were mistreated. The "inhumane conditions" that Jurnove observed at the game park did indeed cause him suffering, as did the USDA's response to his complaints. Jurnove was distressed by the park's failure to provide any psychological enrichment for its monkeys as required by Congress's 1984 AWA amendments. Instead, he observed an isolated Japanese Snow Macaque huddled unhappily in its cage, and a single chimpanzee named Barney longing for social contact with other chimps. Jurnove repeatedly contacted the USDA to secure help for these animals; the USDA did visit the zoo several times, but on each occasion the inspectors "found the facility in compliance with all the standards." From the ALDF's perspective this situation clearly indicated that the USDA had failed to "adopt specific, minimum standards to protect primates' psychological well-being" but had instead "delegated this responsibility to regulated entities by requiring that regulated entities devise 'plans' for this purpose."[35] The court agreed with the ALDF and ruled against the USDA.

Two years later, in *Animal Legal Defense Fund v. Daniel Glickman* (1998), the U.S. Court of Appeals for the District of Columbia rejected the government's argument that the lower court had been wrong to grant standing to the ALDF. The court not only affirmed Jurnove's specific injury, but also held that animal rights groups should be able to "invoke the aid of the courts in enforcing" laws when they observed animals "in a persistent state of suffering" because of governmental inaction. Moreover, the court read the legislative history of the AWA to indicate that Congress intended for a "strong and enlightened public" to protect "small helpless creatures." Citing John Mehrtens's letter about roadside zoos used at the 1970 congressional hearing that amended the AWA to include zoos and aquariums, the court determined that "Congress had placed animal exhibitions within the scope of the AWA" precisely because citizens testified about the inhumane conditions they observed. The court concluded that Congress had intended for humane societies to monitor the work of zoos and expected "concerned animal lovers to ensure that the purposes of the Act were honored."[36] The court's decision represented a victory for animal protectionists against the government bureaucracy responsible for implementing the AWA, but not necessarily a victory against zoos themselves. In fact, the AZA actually shared some of the same concerns that the ADLF had with the USDA's development of primate care regulations, and the court discovered that the USDA had given little attention to the needs of primates as it developed its regulations.

In appealing the district court's decision, the National Association of Biomedical Research (NABR) filed a brief that essentially confirmed the ALDF's charge that the government had allowed the regulated to regulate themselves. The NABR revealed that its members—more than 350 public and private research institutions—had invested significant "human and financial resources in the rulemaking proceeding" that resulted in the regulations under discussion. If the appeals court threw out these regulations, NABR members would suffer financial losses by having to go through the process again.[37] The district court, however, had found the six-year rule-making process to have been unduly influenced by the research industry. The court found little evidence that the agency had taken seriously its responsibility to protect the welfare of primates, observing that it had collected thousands of comments during the process but had not used these comments to *create* the standards published in 1991. Instead, it simply allowed institutions that owned nonhuman primates to set their own minimum standards.[38] The court implicated Congress in the entire fiasco, charging that it had "set forth a clear mandate of humane treatment of animals, [but] it then took away from that mandate by granting unbridled discretion to the agency, which as past experience indicates, will do little or nothing."[39]

The case represented a clear statement of the animal protectionist groups' right to secure third-party standing to protect animals, but it was not a victory for animal protection groups in their battles with the AZA. The district court focused its wrath on the "failures of our system of government," not the failure of the Long Island Game Park and Zoo to create an adequate plan for the humane treatment of its primates. In fact, the court disallowed the introduction of videotapes documenting the condition of the zoo's facilities on the grounds that such evidence would "present an incomplete picture," because the USDA had several thousand facilities under its control and not all of them were necessarily as bad as this one.[40] The AZA historically offered the same line of argumentation whenever animal protectionists publicized a case of animal abuse in zoos. Indeed, the AZA itself felt that the USDA's rules regarding the well-being of nonhuman primates were inadequate. During the 1989 comment period, the AAZPA submitted numerous recommendations and criticisms, including the need for species-specific minimum space requirements and "concern that the proposed regulations did not include a definition of 'psychological well-being' . . . nor a list of criteria by which to judge" whether psychological needs were being met.[41] When APHIS published the final rules, the AAZPA discouragingly reported that APHIS adopted few of their suggestions and did not consider the AAZPA's space recommendations "appropriate to require in regulations because they exceed the minimum space necessary for the humane care of nonhuman primates."[42] By establishing minimum requirements that it knew would be higher than the industry standard, the AZA had reduced the likelihood that someone such as Jurnove would come away from a visit to one of their institutions with the impression that the zoo was doing nothing to promote the well-being of its animals.

Despite animal protectionists' complaints, the AZA zoos gave significant attention to the welfare of animals inside their gates. Although research and agriculture interests tended to lobby government for minimum humane treatment regulations, the AZA consistently sought higher standards. When animal protectionists perceived abuse, as in the transfers of Lota and Timmy, zoos could reasonably argue that they were acting to protect those animals. Even the euthanasia of animals such as the Detroit tigers could be justified as humane, when those animals had been suffering. On the other hand, zoos had much more difficulty defending the transfer and euthanasia of "surplus" animals that had no health problems and posed no risk to other animals. The development of a defensible policy regarding such animals came to preoccupy and divide AZA members during the 1990s.

A Surplus Animal Policy

The AZA enjoyed legal and legislative successes during the 1990s, but it waged a much more difficult war internally and in the public relations arena over the continuing problem of how zoos disposed of their "surplus" animals. Zoos could proudly publicize their high standards of animal care, their conservation activities, and even their successful breeding of endangered species, but they preferred to keep silent about the fact that some of their unwanted animals ended up dead or in the hands of incompetent caretakers. Animal protectionists took every opportunity, however, to continue publicizing this kind of animal "abuse" by zoos. Worse, concerned zoo employees were going public about this dark side of the zoo, and the AZA membership itself was divided about how to deal with the issue. Slowly, the AZA developed a surplus animal policy, and by the end of the decade it had begun to build some public credibility by lending visible support to efforts to restrict the private ownership—and, by extension, abuse—of exotic pets.

As the decade opened, zoos again faced external and internal pressure to address the surplus animal issue. Animal rights groups began staging protests in front of zoos and aquariums and passing out leaflets to build public awareness about the surplus problem. The Friends of Animals, for example, sent the Oklahoma City Zoo a leaflet titled "Zeroing in on Zoos" that it distributed outside of the zoos it picketed. All of its arguments against zoos focused on the disposal of unwanted animals. The leaflet stated dramatically that "almost every major zoo in the country is either contributing to the problem or turning its back on it," and it alleged that even the prestigious San Diego Zoological Society sent a Dybowski's sika deer to a hunting ranch. In addition to confronting protesters at their gates, zoos also faced internal criticism from credible sources. In 1991, Donald Lindburg, the editor-in-chief of *Zoo*

Biology, wrote an editorial critical of the zoo surplus animal problem, which Wagner distributed to the AAZPA board of directors. It became difficult for the AAZPA to ignore the fact that breeding programs were producing more animals than zoos could exhibit and that the surplus animals were sometimes being euthanized or sold to exotic animal dealers, eventually ending up in the hands of private owners, roadside zoos, or even hunting ranches. The AAZPA had tackled the surplus issue in the 1970s and again in the 1980s, but it appeared to be a problem with no easy solution.[43]

To address this growing problem, the AAZPA formed a surplus animal fact-finding committee in 1990. The committee's resulting report focused on the place of euthanasia in the dealing with unwanted animals. The report illustrated at least some of the AAZPA's understanding of when euthanasia should be employed, the uncertainty zoo members themselves felt about the issue, what they thought they should do about the problem, the trouble that zoos members had sympathizing with animal protection groups on the issue, and their strategy for managing the public relations difficulties that accompanied killing zoo animals.

The report confirmed that zoo animals were in fact ending up as pets or on hunting ranches. Using ISIS data (an animal registry system for zoos), the authors calculated the number of animals "removed" from zoos. Although their estimates were rough, because not all AAZPA accredited institutions participated in the ISIS system, they indicated that as many as 5 percent of all zoo animals were removed from their homes each year. Most of these animals went to other zoos, but the authors concluded that a significant number of animals ended up in the hands of private dealers and individuals. On the basis of their findings, they made suggestions for surplus animal guidelines that included increased education about the issue and an agreement between each zoo and those who took their animals. This agreement would control what happened to the animals in subsequent transactions by prohibiting the new owner from selling them to an inhumane research program, allowing them to be hunted, or selling them to people who were suspected of animal abuse. The report also offered recommendations about how zoos could keep most animals out of the hands of private citizens and hunting institutions: use birth control, separate the sexes, give the animals to another qualified zoo, sell to an accredited dealer, or give them to regulatory agencies for reintroduction. Although zoos hoped to send some animals back to the wild, the authors did not anticipate being able to do this often for "the next century or two." There were a few animals that fell into a "gray area" between pets and wildlife that the report indicated might be confidently sold back to the public through reputable animal dealers. As a last resort the report recommended using euthanasia, the most controversial method of animal disposal.[44]

The report suggested conditions under which euthanasia might be employed. Notably absent was the condition of an aggressive temperament, as had been the case for one of the Detroit Zoo's Siberian tigers. Instead, the recommendations centered on poor health and population management. For example, the report approved of euthanasia in the cases where "animals receiving medical attention do not respond to treatment," "animals cannot carry out minimal biological functions," or "animals [have] no realistic chance of survival." At the same time, it reminded its readers that a commitment to saving species required preserving the gene pools, which had to be "managed" so that the surplus animals whose genes were redundant did not "deprive" other animals of a place on the "captive-ark." At least some of these euthanized animals might make appropriate food for other animals in the facility.[45]

The internal political problem, the report acknowledged, was that not everyone within a given zoo supported euthanasia. The report noted that, because zoo keepers often developed an "emotional rapport" with the individual animals for which they cared, they were particularly reluctant to approve of euthanasia. The report noted that keepers agreed with euthanasia in theory as a means to "manage genetic diversity," but they often objected to it in practice. As a result, the report recommended educating both keepers and volunteers whose "sentimental involvement may be even more of a motivation" for their job.[46] As the Detroit tiger case showed, unhappy zoo employees were more than a hypothetical possibility.

In addition to identifying problems with zoos' own employees, the report detailed the public's substantial resistance to euthanasia. Zoos unintentionally heightened the public's emotional feelings about animals with their "adopt an animal" fund-raising programs in the 1980s that encouraged citizens to believe that they owned a particular animal. Thus, the report recommended doing away with these programs, "de-emphasizing individual animals and . . . addressing species as a whole." Doing so would help the public, which the report described as "lack[ing] information and understanding of animals," accept euthanasia. Animal enthusiasts, the report continued, had "limited intellectual and ecological understanding of animals, with a very high humanistic attitude."[47]

In addition to taking away animals' names and separating donors from animals, the report also recommended an elaborate plan to manage the potential public relations disaster lurking in every euthanasia decision. It advised careful documentation of why a particular animal was "surplus" through reference to its genetic redundancy. Following that, it suggested gathering keepers and other zoo professionals together for a meeting and handing out the AAZPA surplus guidelines and other reference materials on euthanasia. To head off criticism from public authorities such as city councils, which often had governing authority over the animals, the report recommended "stressing the risks of disposing . . . surplus animals to unqualified recipients and the negative long-range effects of excessive birth control upon the survival of endangered species."[48] In short, the report recognized, as the judge in Detroit's tiger case had opined, that euthanasia decisions were ultimately political.

As public institutions, zoos would have to generate public understanding of their policies, yet the issues related to surplus animals remained contentious, even within the zoo community. These divisions were clearly visible at a 1993 AAZPA forum on surplus animals and hunting. On the one hand, some members resisted any accommodation on the issue, defending zoos' right to dispose of animals in any manner they saw fit, including sales to hunting ranches. One member noted that zoos were regulated by the USDA and wondered, "why should zoo animals be legally considered different from any other form of livestock?" A more moderate voice put the issue in the context of political attacks on zoos, arguing that zoos should "seek a position that would provide for a management policy based upon conservation principles, rather than . . . one which appears merely to serve the animal rights agenda." Speakers on this side expressed unease about maintaining responsibility for animals once they left a zoo's gates. On the other hand, some members insisted that zoos should care about the fate of all animals, not just those in their immediate care. The antihunting faction within the AAZPA argued that the AAZPA's philosophical support for the sustainable harvesting of natural resources did not include "taking a zoo-raised animal, putting it in a crate and allowing someone to shoot it as it is released." Others reminded their fellow members that zoos gained little political benefit from supporting game ranches: "why does the AAZPA want to be associated with these people?"[49]

These antihunting arguments were compelling to the AAZPA board, which apparently did not want to be associated with "those people" either. Although at least some of its members characterized animal enthusiasts as lacking in intelligence, the AAZPA board ultimately came down on the side of those members who wanted to protect individual animals. In its policy statement about the disposition of animals to hunting organizations, it reminded its members that zoo animals are "held in public trust" by largely public, taxpayer-supported institutions and that the public certainly did not imagine that its zoos were breeding animals for big-game hunters. Though they noted that some conservation policy involved culling, they stated that sending wildlife to hunting ranches impugned the role of zoos "as sensitive guardians and conservators."[50] Just as zoos' public status guaranteed them some legal protections, it also obligated them to be at least somewhat responsive to popular opinion.

Ultimately the AAZPA was able to resolve the surplus issue as it related to hunting ranches and embarked on a campaign with animal protection groups aimed at curtailing exotic pet ownership. The HSUS, meanwhile, had embarked on a surplus animal saving program of its own.

The Sugarloaf Dolphin Sanctuary

Zoos were not the only institutions grappling with a surplus animal problem in the 1990s. With the end of the cold war, the Navy was encouraged by legislators such as Senator Robert Byrd (D-WV) and Representative Charles Wilson (D-TX) and animal protection groups such as the HSUS to relocate most, if not all, of its one hundred dolphins housed in San Diego under the care of the Marine Mammal Systems program in the Space and Naval Warfare system. The dolphins were "cold warriors," captured off the coast of Florida by a private company, Marine Mammal Productions, for the Navy, whose trainers, such as Kerry Sullivan, taught them a variety of tricks for their health care needs and military use. The sudden availability of dolphins, however, worried the HSUS, which did not want to see these animals distributed to zoos and aquariums. At the time, the HSUS was involved in the lawsuit against the Shedd Aquarium for importing dolphins for their exhibit. Dr. Naomi Rose, the HSUS's marine mammal scientist, offered to prepare the Navy dolphins for reintroduction along the lines that zoos had been experimenting with for years. One HSUS member, Karen Hode, offered to finance the rehabilitation and release program. This program would require the HSUS to implement *practically* its vision of returning the animals to the wild. The HSUS, however, needed a dolphin-holding facility for the rehabilitation work.

In the meantime, O'Barry also knew about the Navy dolphins that needed a home. Lacking a facility of his own, O'Barry teamed up with Lloyd Good III, the owner of Florida's Sugarloaf Lodge resort, whose family had had a pet dolphin named Sugar for around twenty years. Sugar was kept in a pen in the lagoon adjacent to the resort. O'Barry and Good planned to train three Navy dolphins—Luther, Buck, and Jake—for reintroduction into the wild while they also trained the Ocean Reef resort's Bogie and Bacall for release. Rose, O'Barry, and Good had selected these dolphins as the best possible candidates from the Navy's San Diego facility and had them flown to the sanctuary. O'Barry had previously rescued a dolphin from a Brazil amusement park, where it showed signs of serious mental deterioration, including repeated knocking of its head against the wall and tearing off portions of its flesh. He had also appeared in nature documentaries and testified before Congress as a dolphin rights activists and therefore seemed to be a good contact for the HSUS to work with in releasing at least a few Navy dolphins.

Congressman Wilson set up a meeting between Rose, the HSUS's vice president of government affairs, Wayne Biselli, and Rear Admiral Walter Cantrell, who had the authority to authorize the release and was eager to facilitate a transfer of dolphins. The HSUS agreed to help finance the dolphin release, facilitate the permitting process, and act as a liaison with the Navy. According to Rose they were not "just interested in releasing the dolphins," but also wanted to do it "in such as way that the public display industry could find no fault with it."[51]

To obtain the dolphins, O'Barry and Good required two permits. They needed a public display permit under the AWA that allowed them to house the dolphins and assured APHIS that they had a proper veterinarian, a secure enclosure for the dolphins' safety, and the necessary nutrition. They also needed an MMPA scientific research permit. Scientific research permits effectively prevent the holder from being held liable if the animals are hurt in the process of a legitimate study. To receive the permit, Good and O'Barry needed to write down how they were going to rehabilitate and release the dolphins—the "protocol"—to the public and a group of marine mammal scientists, all of whom could comment on the plans to make sure that O'Barry and Good had a plan with a good chance of success. Reintroduction of longtime-captive dolphins is not simply a case of releasing the animals into the ocean. The dolphins' trained dependence on people must be erased, they must be able to capture live fish, and they must be with other animals of the same stock so that zoos and aquariums do not inadvertently create new species by releasing, for example, an Atlantic bottlenose dolphin into the Pacific. The research permit must also show how the people plan to prevent illnesses specific to marine mammals in captivity from being introduced into wild marine mammal populations. And, perhaps most difficult, the permit holder must show that the animals in captivity will not change the behavior of those in the wild by demonstrating behaviors that are only found in captivity. Dolphins, for example, learn by imitation, and one does not want to teach pods of dolphins to beg for fish from boats, as captive dolphins will do. Making this task all the more challenging was the fact that in the 1990s there had been only two scientifically documented dolphin reintroductions, one of which was conducted by Dr. Randall Wells, a biologist from the Chicago Zoological Society, which governs the Brookfield Zoo. Wells had retrained two dolphins—Misha and Echo—who were captured in the Tampa Bay area in the 1980s and given to Professor Ken Norris at the University of California—Santa Cruz to research the echo-location abilities of dolphins. When the research was completed, Wells retrained the animals to catch live fish, ended their connections with humans, and monitored them closely by boat for a year. He documented his work and published it in *Marine Mammal Science*.[52]

O'Barry and Good, with the help of Rose, were supposed to engage in the same process. Initially their work seemed promising. In 1994, Good asked directors of established marine mammal conservancy organizations to serve on the board of directors of what became known as the Sugarloaf Dolphin Sanctuary. The board of directors included Rick Trout, from the Marine Mammal Conservancy in Florida; Rick Spill, a self-described dolphin activist; Robert Schoelkopf, the cofounder of the New Jersey–based Marine Mammal Stranding Center; and Mark Berman, from the Earth Island Institute, which was currently involved in rehabilitating and releasing Keiko, the killer whale portrayed in the movie *Free Willy*. In 1994, Spill organized a "gadflies coalition" meeting of dolphin rights activists that O'Barry and Good hosted at Sugarloaf. The meeting brought together activists who believed that "the display of marine mammals, especially whales and dolphins, was wrong." Rose attended the meeting and met her future inside informant Kathy Kinsman, another dolphin activist, who was making a music video. The auspicious start not withstanding, the work at the sanctuary quickly turned chaotic.[53]

O'Barry and Good applied for a scientific research permit, but neither man was in favor of that kind of approach to releasing the dolphins. O'Barry dictated his version of a release protocol, but it was "an exercise in Zen" so there were no possible protocols because it was "completely non-verbal." The idea was to lose oneself and "become one with the dolphin." He hoped to "pitch a tent" near the dolphins and become "part of the scenery, like one of the trees or heron." Good's idea for releasing the dolphins involved simply letting them decide when they were ready to leave. Meanwhile, the dolphins were also occasionally swimming with the guests and fed by them. Good and O'Barry flung live fish to the dolphins while hiding behind a tree so that the dolphins would disassociate food and humans, but the experiment went awry as nearby seagulls sensed a free meal and began attacking the dolphins' heads while they waited for the fish to fall from the sky. Fearing that the entire release program was in jeopardy, the HSUS first tried to have Mary Lycan, a local dolphin trainer, take over writing the release protocol for the research permit. Good and O'Barry, however, believed that they were being supplanted by the HSUS in their own sanctuary and ultimately banned Lycan from the facility. The HSUS was being kept informed about the deteriorating situation at the sanctuary by Kathy Kinsman, who was calling Rose to keep her informed. Recognizing that they were losing control of the dolphins, the HSUS severed their relationship with the sanctuary in 1995, but continued to get inside information from Kathy Kinsman.[54]

Veterinarians also struggled to work with Good at the sanctuary. The problem was that Good had a holistic approach to dolphin health that did not involve invasive medical testing, such as drawing blood. A string of veterinarians argued with Good but ultimately left, failing to receive sufficient access to the animals. Eventually, Good asked his retriever's veterinarian to serve as the dolphins' doctor because he needed a veterinarian to keep his APHIS permit. This failed to satisfy APHIS, however, because Good's veterinarian had received his degree from the Dominican Republic, which the state of Florida refused to recognize.

The Sugarloaf Dolphin Sanctuary's board members held an emergency meeting in June 1995 at which they drafted a letter to Good and O'Barry alleging that water-quality tests at the lagoon had been bad, that Jake needed better veterinary treatment for lesions, and that the management of Sugarloaf "ranged from poor to dismal, including a largely absent director, unfounded claims of coups, continued questionable firings," and "indications of possible monetary improprieties." Angered at the accusations, O'Barry and Good dismissed their fellow board members and took complete control of the sanctuary. In direct violation of their AWA permit, Good let the dolphins out of their enclosure to explore the lagoon and, he hoped, learn that they could survive on their own. Fearing for the dolphins' safety in open water, O'Barry alerted APHIS that Good was violating the terms of their AWA permit. By 1996 the National Marine Fisheries Service (NMFS) was poised to confiscate the dolphins when someone inside tipped off O'Barry and Good that they were coming. Concerned that the dolphins might go back to the Navy, O'Barry and Good rented a boat, took Luther and Buck six miles off the Keys and set them free, a process documented by a French film.[55]

Neither Luther nor Buck, however, was ready for release. Instead of joining nearby pods of dolphins, they began swimming near boats in marinas. Rather than catching fresh fish, they were opening their mouths and waiting to be fed by passing boats. Luther, who like Buck was marked with a freeze brand, hung around the Sunset Marina on Memorial Day, causing boat traffic jams as vacationers slowed down to observe, feed, and swim with him. Citizens reported the friendly dolphins to the Dolphin Research Center, a local nonprofit research and educational center that began searching for the animals. With the help of NMFS, the Marine Mammal Conservancy, and the Navy, the dolphins were recaptured and given treatment for injuries and being underweight.

The National Oceanic and Atmospheric Administration initiated a civil penalty against O'Barry and Good for seven violations of the MMPA in 1997, and the trial was eventually held in June 1999 in the Keys before Peter Fitzpatrick, an administrative law judge. The court found both Good and O'Barry guilty of violating the MMPA on several counts because they had "harassed" the dolphins by putting their lives in harm's way and had failed to obtain a scientific research permit. Testifying as expert witnesses at the trial, veterinarians noted that both dolphins had cuts on their sides, one of which was likely caused by a boat propeller. Buck was also clearly underweight when captured and had symptoms of "nutritional compromise," according to a veterinarian from the University of Miami who examined him.[56]

During their trial O'Barry and Good took full responsibility for their behavior, admitting that they made a regrettable decision but that it was an "act of civil disobedience" designed partly to keep the dolphins away from the Navy. Lending some credibility to their fears was the fact that Jake, who had been confiscated and returned to the Navy for safe keeping, died in the Navy's care in February of 1999 from excessive fluid in his stomach and, ultimately, cardiac arrest. At the trial, Judge Fitzpatrick also expressed his dismay at the Navy's lack of concern about the dolphins. He read a portion of a newspaper article to Dr. William Van Bonn, the Navy's dolphin veterinarian, in which a Navy representative commented on the death by saying, "we are sorry he is dead, but that's the way it goes." The judge asked Van Bonn whether he thought this was "an amazingly callous statement." O'Barry and Good also reminded the judge and the witnesses about the Navy's Vietnam War–era dolphin swimmer nullification program, in which dolphins wore devices strapped to their heads that could discharge and kill a diver. The shock of the explosion, however, sometimes broke the dolphins' jaws.[57]

In addition to fearing that the dolphins would go back to the Navy, O'Barry felt that it was also "his only choice left" after experiencing "intense agency interference" in which APHIS inspected his facility forty-one times in two and a half years. He believed that he was "a victim of a carefully orchestrated strategy by both National Marine Fisheries Service and the USDA." Explaining the lack of support by fellow animal protection groups, he argued that the HSUS failed to support him because they were "rival activists" who "attached themselves to this product." He and Good noted that Spill was actually using an alias—that his real name was Rick Wewer. O'Barry alleged that he was really a "plant sent there to destroy the project" by the group Putting People First, an anti–animal rights group. Similarly, Good explained Trout's failure to support the sanctuary by noting that he was a "dolphin rights activist; and he attacks things." Because he "couldn't switch gears," he began arguing against their sanctuary as well. Good felt that he was caught in the middle of both the political left and right. On the "extreme right" was the "captive display industry," which included the "Dolphin Research Center," an institution Good alleged was "threatened by the idea of the gate being opened" fearing that "the dolphins might actually decide to leave," setting a "bad precedent" by "threatening their livelihood." On the far left, apparently, was O'Barry, who "makes his living" from documentary films and interest group monetary support by "saying that captivity is bad and returning [dolphins] to the wild is good." Good alleged that O'Barry was "equally freaked out that the gate was open" because he was "afraid that the dolphins might stay" in the sanctuary and thereby, Good implied, threaten O'Barry's means of supporting himself financially.[58]

The court concluded that the release was an "absolute tragedy" because it was done so poorly. Fitzpatrick opined that, because of this debacle, the "whole process of release" would have to be "carefully looked at to avoid the very kinds of things that happened right here." The judge wondered, in fact, whether they had "placed in issue" the whole noble idea of dolphin reintroductions. He wondered what the Navy should do now. To whom should they give their dolphins? "What about NMFS? What about the USDA? Who can they trust now?" the judge questioned.[59]

The answer to these questions was in the case. The Chicago Zoological Society's Wells had successfully obtained the necessary permit and released two dolphins back into the wild. The court's focus on O'Barry and Good, while necessary, obscured the substantially flawed involvement of the HSUS in facilitating the dolphin transfer to the sanctuary. By Rose's own admission, Congressman Wilson refused to make the transfer without the assistance of the HSUS, "a large credible organization." At that point, the HSUS could have insisted that they work with Wells, but, because they opposed aquariums, they refused to do so and instead elected to work with well-intentioned fellow activists with no scientific credentials. The HSUS's inability to carry out a successful reintroduction seemed to confirm what AAZPA president David Zucconi had explained in a more general way to the HSUS's John Grandy in 1990: organizations such as the HSUS were quite effective at calling for change in animal welfare practices, but only the AAZPA had the qualifications to actually care for exotic animals.[60]

Conclusion

In several respects, the zoo community appeared to have strengthened its position in the political system during the 1990s. The AZA had matured as a political organization. By renaming itself with a simpler acronym and reorienting its publications toward the general public, the association had moved beyond serving its members in a private capacity to establish a more visible image that would benefit the zoo community in its public debates with anti-zoo forces. The association had also solidified zoos' conservation image by selecting a new executive director with unquestionable conservation credentials. On the legislative side, zoos not only continued to support environmental initiatives, but also gained important legal protections from radical animal rights' activists. In addition, for-profit AZA member institutions benefited from further affirmations of their right to import endangered animals such as giant pandas.

By any objective measure, zoo animals also enjoyed greater protections by the end of the 1990s than they had in the early 1970s. The AWA, ESA, and MMPA remained in full force, and, though the AZA had succeeded in shaping their implementation to serve zoos' interests, the movement and display of zoo animals continued to be covered by all three laws. The zoo community itself had made dramatic improvements in its care of animals. Zoos built larger, naturalistic enclosures for their animals, hired veterinarians, and professionalized animal care. An increased focus on research enabled zoo professionals to expand their knowledge of captive animals' dietary and behavioral requirements greatly. Even surplus animals, which received little thought from most zoo professionals in the early 1970s, had become the subject of much concern and debate.

These improvements in zoo animals' welfare, however, failed to satisfy animal protection groups, which continued to battle zoos in the courts, pushing the questions of whether zoos were commercial institutions and whether animals had rights akin to humans. The potential danger of these groups was evident in their occasional physical attacks on zoo and aquarium property. Perhaps more troubling, the HSUS, previously a somewhat cooperative animal welfare organization, adopted a more radical position on zoos, condemning them regardless of the quality of their facilities. This position did not clearly benefit animals, as seen when the HSUS turned to fellow activists at a hastily constructed dolphin sanctuary to implement their larger goal of reintroducing captive animals into the wild. The Sugarloaf Dolphin Sanctuary experience gave the HSUS a taste of how difficult it was to work with and release animals rather than solely engage in political agitation for animal welfare.

Optimists within the AZA had once believed that zoo professionals and animal protectionists could work together on many issues, and indeed episodes of cooperation did continue through the 1990s. At a deeper level, however, events in the 1990s illustrated that the zoo professional and animal welfare activists held increasingly different definitions of "protection." Because of their scientific and conservation orientation, zoo professionals often sought the welfare of animal species above that of individual animals. By contrast, animal welfare activists, influenced by animal rights philosophers, focused greater attention on the welfare of individual captive animals. These different philosophical perspectives made cooperation between zoos and animal welfare groups more difficult. The AZA had chartered a course that focused on managing captive species, and "management" required the culling of old and genetically unfit animals. Zoos struggled to find a solution to this problem, but none really existed. The only sure way to prevent surplus animals was to stop breeding zoo animals, but that solution would mean the eventual death of zoos. The logic of the animal protectionists' argument appeared to point in this direction.

How could the zoo community best protect both wild and captive animals? It had long sought to protect wild animals outside zoo gates, lending support to environmental legislation, funding conservation activities, reintroducing captive bred animals into the wild, and educating visitors about the plight of wild animals. As Conway wrote to a colleague in 1981, zoos should "become the leading centers of environmental education, the human institutions which ultimately represent, embody and teach environmentalism."[61] In many respects, the zoo community could feel proud about its efforts on behalf of wild animals. Nevertheless, the zoo community still struggled with the question of whether or how to protect their own animals once they passed beyond a zoo's gates. It appeared that the plight of surplus zoo animals would remain a vulnerability exploited by animal protectionists.

The zoo community thus entered the twenty-first century with a mixed sense of relief and foreboding. In a very real sense, zoos had survived and had even been strengthened by three decades of legislation. Numerous court decisions had affirmed that the AWA, ESA, and MMPA contained no teeth with which animal protectionists could substantially harm zoos or aquariums. The AZA had significantly professionalized the zoo community, ensuring that captive animals received better care than at any time in zoos' history. But the animal protectionist groups showed no signs of going away, and if anything their tactics and rhetoric were becoming more antizoo. These groups had seemingly exhausted legal challenges. What tactics would they employ next?

CONCLUSION

Elephants and the Trajectory of Zoo Politics

Many of the issues that bedeviled zoos and aquariums between the 1960s and 1990s remain today. Animal protection groups continue to question the welfare of zoo animals and pursue tested political strategies. Whereas marine mammals drew much attention from protectionists between the 1970s and 1990s, elephants have become the new target. Inspired in part by research detailing health problems in captive elephants, animal welfare and rights groups have joined forces to remove elephants from zoos. This latest challenge for the zoo community reminds us that, though the specifics may change, a familiar set of themes continues to animate zoo politics. In confronting the real possibility that animal protectionists might make it politically difficult for zoos to exhibit elephants, the AZA leadership once again has had to articulate the zoo community's public mission, craft a political strategy for a new environment, seek alliances with protectionists and politicians alike, and impose unity on a membership divided over the elephant question.

Protectionists have focused on elephants for a number of reasons. First, activists abroad have placed elephants at the center of zoo debates. A 2002 study of European zoos commissioned by the English Royal Society for Prevention of Cruelty to Animals claimed that zoo elephants suffered from a variety of health ailments. This study led to calls for the British zoo community to phase out elephants from its collections, potentially setting a precedent that protectionists would like to see followed in the United States. Second, the American zoo community has internally debated the proper training, treatment, and use of elephants for decades. Finally, alleged elephant abuses by circuses have implicated zoos and made them vulnerable to protectionists' criticisms. Because circuses are privately owned, commercial operations with a primary focus on using animals for entertainment purposes, it is more difficult for them to claim that they are educational. Yet, as demonstrated in earlier chapters, zoos sometimes transfer surplus elephants to circuses, to the peril of both elephant and zoo.

After Lota, the Milwaukee Zoo elephant, went to the Hawthorn Corporation, she contracted tuberculosis. Under the scrutiny of animal protection groups, the U.S. Department of Agriculture (USDA) investigated and charged the company and its president, John Cuneo, with violating the Animal Welfare Act (AWA). The Milwaukee Zoological Society, meanwhile, set aside $20,000 to "help fund Lota's removal from Cuneo's facility" to "anywhere that she'll receive a high level of care and where she can live out her life." Once it became clear that the elephant was not thriving in Hawthorn's care, the society considered legal means to remove her but was unable to do so. Eventually Cuneo admitted guilt on nineteen charges and paid a $200,000 fine, and in 2004 Lota was removed to an elephant sanctuary in Tennessee, where she died three months later from the effects of tuberculosis.[1] The Milwaukee Zoo might have saved itself a considerable amount of money and bad publicity by simply engaging the animal welfare activists more openly from the start and acceding to their wishes, which had always included sending Lota to a sanctuary.

Pursuing the strategy of legal action, protests, and publicity that eventually forced Lota's relocation, animal protection groups are trying to move other elephants from zoos around the country to sanctuaries as well. Beginning in 2004, activists from the Humane Society of the United States (HSUS), In Defense of Animals, People for the Ethical Treatment of Animals (PETA), Save Elephants in Zoos, and other local interest groups launched aggressive campaigns to remove elephants from AZA zoos in Portland, Oregon; Los Angeles; Houston; El Paso; Chicago (Lincoln Park Zoo); and elsewhere. The protectionists argue that elephants require much more land to roam—five acres or more each—and that small exhibits give them potentially lethal foot problems. In addition, they argue that elephants are not suited for cold temperatures and that their psychological health suffers unless they are in large social groups. Activists propose that, if zoos are unable to increase the acreage available for elephants, they ought to send them to large sanctuaries in warmer climates.[2]

Protectionists enjoy some political support for their position. During his 2005 Los Angeles mayoral campaign, candidate Antonio Villaraigosa agreed that the elephants in his city's zoo should be sent to a sanctuary because they lacked sufficient space. Soon after his election, Villaraigosa asked for an "independent review" of the Los Angeles Zoo's two-acre elephant exhibit to determine whether that size was adequate for five elephants, whether the exhibit should be enlarged, or whether the elephants should be sent to a sanctuary. Villaraigosa's public comments indicated that he preferred the final option, and animal activists rejoiced.

After the deaths of two Lincoln Park Zoo elephants that activists had wanted to move to a sanctuary, Chicago alderman George A. Cardenas introduced a nonbinding resolution to move the remaining elephant to a sanctuary and permanently closed the zoo's elephant exhibit. When the third elephant died in May 2005, Alderman Mary Ann Smith proposed an ordinance requiring all elephant exhibitors, including circuses, to provide ten acres for each animal, a condition that the urban Lincoln Park Zoo would not be able to meet. Although Smith's proposal cheered local activists, it appeared during testimony on the issue that some of her fellow council members sided with the zoo community in seeing a significant public value in the exhibition of elephants, and discussions about the exhibition restrictions dragged on through the fall.[3]

As we have seen, in the past the zoo community has used alliances of various sorts to strengthen their side of the debate. Because the elephant issue is being played out at the local level, zoos have turned to city and county politicians. In the Los Angeles mayoral race, the zoo had one candidate firmly in its camp. In El Paso, an AZA representative held a parking lot press conference prior to a city council meeting on the question of moving two city zoo elephants to a sanctuary, and the zoo's supporters spent weeks lining up support from council members. Their efforts helped persuade the council to vote unanimously to keep the elephants at the El Paso Zoo.[4] On the other hand, zoo directors who wish to move their elephants to sanctuaries have enlisted the assistance of animal protectionists, just as the AZA itself has occasionally worked with protectionists and against individual zoos in the interests of protecting animals.

Activists are finding some directors more receptive to their message than others. In 2004, Ron Kagan, the director of the Detroit Zoo, decided to send his Asian elephants Wanda and Winky to a 2,700-acre Tennessee sanctuary operated by a former circus elephant trainer. As Kagan explained in a special brochure titled "Giving Our Elephants Room to Roam" distributed at the zoo, he and his staff worried about the health of the animals in Detroit's cold climate, where they had little space to roam and no contact with a larger social group. Kagan made his decision openly and with the support of city residents. Justifying the move, he argued that, when zoos state they are going to "save wildlife," they are implying that they do not "harm wildlife." He and the elephant keepers at the Detroit Zoo believed that the elephants' quality of life was suffering, which was a problem for animals with "complex minds and probably deep hearts." In April 2005, after several delays resulting from AZA challenges and questions about the suitability of the Tennessee sanctuary, Wanda and Winky were sent to the 2,300-acre Ark 2000 Sanctuary, in San Andreas, California, owned by the Performing Animal Welfare Society (PAWS).[5]

Across the country, the San Francisco Zoo also agreed, under internal and external pressure, to transfer its African elephant Lulu and Asian elephant Tinkerbelle to the Ark 2000 Sanctuary. The zoo had failed to raise sufficient money to improve its small, outdated elephant exhibit and had euthanized Tinkerbelle's partner, Calle, after an accident. Depressed by Calle's death, Tinkerbelle also suffered from feet crippled by thirty-six years of living in a cramped exhibit space and was prone to occasional aggressive behavior. A fourth elephant had died earlier in 2004, and even the zoo's new director, Manuel Mollinedo, believed that Lulu and Tinkerbelle would be better off in a sanctuary until a larger elephant exhibit could be built.[6]

In both San Francisco and Detroit, zoo employees and animal welfare activists supported the idea of placing the elephants in a sanctuary. Tinkerbelle's keeper, Peggy Farr, noticed a huge improvement in her attitude after just a few days at the PAWS sanctuary and said that it was "nice to see her up here," though four months later, on March 24, 2005, Tinkerbelle collapsed and was euthanized. Animal rights activists in both cities supported the zoos, with the Michigan Humane Society praising the Detroit Zoo for its "visionary leadership." In San Francisco, animal rights activists convinced the board of supervisors to consider a total ban on elephants. Detroit activists made no similar political inroads, but Gary Yourofsky, founder of Animals Deserve Absolute Protection Today and Tomorrow, hoped that the remainder of the zoo's animals would eventually follow the elephants and be replaced with "virtual" wildlife exhibits.[7]

Yourofky's actual agenda was not missed by the AZA, which recognized the deeper danger implicit in the protectionists' strategy of "saving" a single animal at a time. Americans are certainly supportive of providing the best care possible for ailing animals, but their consistently high attendance at zoos indicates that they do not desire to replace zoos with television shows. For the sake of its member institutions and the public that supports zoos, the AZA went on the offensive against the elephant removal strategy.

Initially, the AZA tried to create a unified zoo community response by threatening to discipline the Detroit and San Francisco zoos if they transferred their elephants to non-AZA institutions. Although endangered, none of the elephants required Endangered Species Act (ESA) permits for transfer because they were born before 1973 and were not going to be used for commercial purposes. Unable to use the force of law, the AZA instead argued that Wanda and Winky were needed for the elephant Species Survival Plan (SSP). The AZA expected its member institutions to abide by the rules of SSPs, which stress not just the breeding of endangered animals, but also their placement in other AZA institutions. Sending the elephants to a sanctuary without first offering them to other zoos appeared to be an attempt to "circumvent AZA conservation programs" and could be "detrimental to the species involved." In the Detroit case, the AZA wanted the elephants to go to the Columbus (Ohio) Zoo, a move that Kagan rejected on the grounds that the Columbus elephant facility was not equal to the sanctuary in terms of what it could offer Wanda and Winky. In statements to the press, the AZA's conservation director, Michael Hutchins, argued that, by attracting the public to zoos, elephants played an important role in conservation efforts. This dispute between conservation and welfare perspectives ended with the AZA saving face when the elephant SSP declared the Detroit elephants "non-essential to the population" on the grounds that Wanda might carry the elephant endotheliotropic herpesvirus, which would endanger breeding herds. The threat of losing its accreditation still hung over the San Francisco Zoo, however, because the AZA suspected that political pressure "by the Board of Supervisors or other such authorities," not the zoo director's best judgment, led to the elephant transfer decision. Ultimately, however, the fact that no federal laws governed the movement of these specific elephants greatly reduced the AZA's sanctioning power. It could threaten the zoos with lost accreditation, but it could not point to federal laws that would justify that action, as it had in the Toledo panda case. Moreover, both zoos enjoyed clear local support for their decisions, and a prolonged battle by the AZA would only damage its image as an organization supporting facilities "dedicated to providing excellent care for animals," as it remarks at the end of its press releases.[8]

In some respects, the willingness of the San Francisco and Detroit zoos to go against the AZA's wishes resembled the Toledo Zoo panda controversy and reminds us that the AZA leadership has continued to struggle in large and small ways to maintain member unity. In the elephant cases, the AZA abandoned attempts to coerce unity from its members and turned instead to persuasion. It reacted quickly, holding a series of strategic planning meetings for directors whose institutions held elephants. The purpose of these meetings was to develop a unified plan of response to the increasing pressure revolving around elephants. The meeting participants agreed that in the future all "AZA elephant facilities will speak with a unified external voice," and they drafted a statement—"Vision: AZA Elephant Management Program"—for board approval.[9] The AZA emphasized central control and member unity in the management of the captive elephant population precisely because the entire zoo community could be affected negatively by the actions of individual zoos. In the past, the AZA worried about member institutions with inadequate animal care standards. In the elephant case, it sought to prevent individual zoos from caring too much about individual animals at the expense of the captive population.

The AZA was also following its tradition of employing flexible political strategies to meet new challenges. In this case, the AZA engaged in an aggressive public relations offensive designed to keep Americans satisfied that zoos cared well for their elephants. The association commissioned a Harris online survey about elephants and publicized the results widely, which showed strong support for the continued exhibition of elephants. It also devoted a considerable amount of space on its Web site to answering the charges of animal protection activists in detail. In a *Communiqué* article, Michael Hutchins, director of the AZA's department of conservation and science, made the case for elephants at zoos rather than sanctuaries. Hutchins offered the now familiar arguments comparing the well-established accreditation process of the AZA to the relatively new and less advanced one of sanctuaries. Hinting at the lack of scientific training evident in the Sugarloaf Sanctuary debacle, he stressed that "many people with master's level and Ph.D.s now work in AZA institutions." He pointed out that sanctuaries did not participate in SSP programs because they did not breed the animals. Sanctuaries, unlike zoos, have adopted the singular mission of individual animal care rather than species preservation and have focused simply on helping individual animals live out the rest of their lives in the best health and accommodations possible. The lack of breeding, however, denies the elephants access to one of their important life functions (reproduction) and simultaneously prevents them from creating new social groupings. Hutchins also noted that, though zoo elephants are accessible to the public, sanctuaries "offer exclusive viewing opportunities for those who can pay for it"—namely, wealthy donors. The general public would see sanctuary elephants only through live-action webcams. He questioned whether sanctuaries really provide a healthier environment, pointing out that not all are located in subtropical climates such as those where wild elephants live. And finally, Hutchins reminded us that, though they are not technically commercial institutions, sanctuaries established online gift shops to raise money.[10]

Through the battle over the elephants, the struggle over control of the animals continues. Both sides sincerely feel that they are saving animals from the people who love them too much. If charismatic megafauna remain endangered, as it appears they will today, these political battles will only increase in number. The underlying agenda of animal protectionists at the moment is to send Americans back to their homes to watch animals through documentaries, webcams, and television shows. Animals will still be used for "commercial purposes," but the people making the money will be film companies and sanctuaries, rather than zoo and aquarium employees. The central question that Americans and their political proxies will have to address in the future is whether the costs of having the animals in zoos outweigh the good that they do for the public. What makes this decision so difficult is that the millions of Americans who attend zoos and aquariums clearly believe it is important to see and hear the animals for themselves.

In the controversy over elephant displays, we see the now-familiar strategies of animal protection groups and the AZA. Some of the actors may have changed, and the setting has moved from national to local politics, but the arguments and tactics are largely the same. It is reasonable to predict, moreover, that these kinds of political disputes will continue into the future. The animal protection groups appear to have recognized that they have more power at the local level, where they need to convince only a few key politicians to make decisions about zoo policy, than at the national level, where judges, legislators, and bureaucrats are relatively immune to the activists' political pressure. The protectionists' position is strengthened by what has always been zoos' greatest challenge: finding the resources necessary to provide ever-greater standards of care for their animals. When zoos seek but cannot get the millions of dollars needed to renovate their admittedly inadequate elephant exhibits, they are effectively forced to make a public appeal for sanctuaries. In October 2005, the Philadelphia Zoo announced the abandonment of a $22 million elephant exhibit fund-raising drive and began discussions about where to place their four animals. The Friends of Philly Zoo Elephants, which had opposed the construction of a new exhibit, immediately called for the animals to go to a sanctuary, rather than to another zoo.[11] Philadelphia's experience paralleled what happened in San Francisco, and it is likely to occur elsewhere as exhibits age.

The politics of zoos and aquariums in the United States changed dramatically and quickly during the late twentieth century. A revolution occurred in zoos and aquariums between 1919, when the Guzzi family sued the New York Zoological Society, and 1992, when the Animal Protection Institute tried to prevent the Shedd Aquarium from importing dolphins and beluga whales. What, then, can we learn from this history?

First, American zoos and aquariums have been, and remain, enmeshed in the political system in which they reside. Just like the menageries of the past, zoos today are shaped by politics. Instead of monarchs cementing political alliances or demonstrating their military power by controlling animal collections, we now have interest groups, attorneys, city councils, congressmen, bureaucrats, and citizens debating the exhibition of animals. At the most fundamental level, the struggles at zoos and aquariums are over who gets to control the fate of the animals. In the microcosm of the zoo we see interest group politics played out as various organizations struggle to translate the wishes of their constituents into a coherent position on the public display of animals. The changing animal enclosures so frequently noticed by visitors, such as the elaborate new immersion exhibits, mask much deeper political changes that have happened within zoological parks and aquariums in the United States. It is questionable, for example, whether the very animals exhibited at zoos today would have been there without the political protection afforded by the AZA.

Until the 1970s, zoos were largely outside the federal system and subject to few state and local laws. Nevertheless, the AZA began policing its own members to stop their role in endangering species *before* animal protection groups directed their attention to zoos. Animal protection groups only speeded up the conservation mission that was already under way because of central leadership among members of the AZA. Unhappy with conditions at zoos, animal protection groups tried to get the federal government into the business of regulating zoos. They failed to control the legislation regulating zoos significantly, because the AZA had a core group of politically adept members who successfully lobbied members of Congress and regulators to shape zoo legislation and regulation to their own benefit. Representatives of more elite institutions among zoos and aquariums, including the New York Zoological Society, the National Zoo, the New England Aquarium, and the Shedd Aquarium, performed a disproportionate share of the work that the interest group undertook by serving in key political positions within the AZA, testifying at congressional hearings, meeting with bureaucrats from the USDA and the U.S. Department of the Interior, and submitting amicus briefs on behalf of fellow institutions in court cases. Representatives from these key zoos constituted the center of the interest group, at times working with regulatory agencies and against individual zoos for what they believed was the greater good of their profession.[12] As demonstrated in earlier chapters, however, they had to learn how to negotiate the American political system, and they had to create and implement an accreditation program to appease their critics and exempt zoos from further regulations.

In addition, the AZA was aided by members of Congress from both parties who believed that the federal government ought to protect zoos and aquariums rather than abolish them. This history reveals a central lesson, which is that animal protection groups have an uphill battle in their attempts to regulate zoos and aquariums. As seen earlier, the language relating to zoos and aquariums in all of the major wildlife protection acts is unambiguous: the Convention on International Trade in Endangered Species (CITES), the Marine Mammal Protection Act, and the Endangered Species Act contain language *protecting* the ability of zoos and aquariums to collect small numbers of endangered species from the wild. Moreover, Congress has reaffirmed these protections each time it has reauthorized this legislation. The courts are unlikely to ignore this statutory language simply because animal protection groups are frustrated with what they perceive as legal loopholes. The only exceptions to this will probably be court cases in which the defending zoo was not accredited by the AZA and thus lacked the kind of interest group protection that further strengthens the claims of zoos and aquariums that they serve as educational, rather than purely commercial, institutions.

Although animal protection groups were not as successful as they hoped, the competition between these two kinds of interest groups (the AZA and the animal welfare activists) made zoo politics contentious. It also made zoos and aquariums better places. The AZA may very well have created its strong accreditation program in due time, but the added pressure that the animal protection groups placed on zoos and aquariums to upgrade their inferior facilities forced the AZA to speed up its accreditation program in hopes of forestalling future regulation. And, although the animal protection groups have not achieved a single complete success in court battles with the AZA, the threat of potential litigation has led zoos and aquariums to think about how animal protection groups perceive their work and make adjustments that ultimately work to the benefit of animals.[13] As is clear from this history, however, to a large extent animal protection groups were preaching to the converted, because many of their goals were supported by the AZA leadership. It is easy to imagine a war between zoos and activists, and indeed pubic pronouncements from both camps often encouraged such a perception. In reality, however, zoo professionals and animal protectionists often shared concerns. And, as we have seen, at times they privately discussed particularly contentious issues with one another before and after public confrontations.[14]

While animal protection groups helped the AZA realize its own mission of conservation, they also complicated its task because they were not always ideologically or strategically aligned. Like the AZA itself, they struggled internally over how to address the problems of species conservation. Even single interests groups like the HSUS sometimes sent zoos and the AZA conflicting messages about how to address welfare issues. An example is the problem of surplus animals. Should zoos euthanize tigers to prevent them from suffering, or let them live until they die, regardless of their pain? The HSUS's central position favoring euthanasia seemed consistent on this matter, but the society could not control one of its board members whose interest group, the Fund for Animals, had an entirely different perspective. The internal divisions of opinion within the animal protectionist movement meant that the AZA had to decide which course of action would most effectively alleviate criticism, because animal welfare and rights groups were not always united in their prescription for how best to take care of animals inside and outside zoos.

The political changes that occurred during this history were embedded in the economic landscape of zoos as they raised the funds necessary to conform to the regulatory mandates of the AWA and the expanded mission of the AZA. One of the key lessons from zoo's political history since the 1970s has been that, the more zoos stay focused on their role as public institutions, the safer they are. In many ways, zoos are success stories among urban public institutions. While other urban political institutions such as schools decayed because of lack of fiscal support, zoos remade themselves during precisely this same period through the help of citizen activists in zoological societies who rescued them from their sad conditions and refurbished them. This was not a tale of the more efficient, competitive free market vanquishing the inefficient, tax-and-spend public sector. It was a story of interest groups at the local and national levels who were called upon to help in a variety of political arenas. At the local level, for example, zoological societies, typically run by local elites, were given, without competition, the authority to run the zoos while the public held on to the property and agreed to pay for some of the expenses.

The decision to admit commercial members and the privatization of zoos, however, were accompanied by costs for the AZA. Sometimes privatized zoos offered the best living environments for the animals in their care. At other times, their financial need for gate revenues to fund increasingly elaborate exhibits lured them in the direction of collecting animals that were either critically endangered or that animal protection groups felt particularly strongly about not exhibiting. Although the logic of CITES may seem contorted at times, preventing commercial establishments from decimating wildlife is a crucial goal. The more that zoos must compete with other players in the entertainment market, the more vulnerable their animals become to the necessities of companies that are governed by economic interests. Species preservation even within zoos is, at its best, a moral undertaking and may have a better chance of success when monitored by a larger public than that composed of stockholders in a particular corporation.

Another clear lesson from this history is that the interests of biomedical groups and domestic agriculture have been at odds with those of the AZA. Biomedical researchers performing testing on monkeys were not interested in creating large immersion enclosures for their animals. Rather, they argued for the absolute minimum standards for their animals. Similarly, biomedical researchers and the commercial pet industry were not in a hurry to improve the safety of animal transportation. As Robert Wagner has noted, this posed problems throughout the AZA's history because the AZA "was concerned that [they] were often painted with the same brush as some in the animal industry whose purpose did not always seem to benefit animals."[15] The AZA took advantage of the greater political muscle of biomedical researchers only when it looked like they were going to face increasing physical attacks by animal rights groups. The irony then, is that the actual, and threatened, physical attacks of animal rights groups actually pushed the AZA closer to the very biomedical researchers that animal rights groups disliked the most.

Zoo and aquarium professionals are caught to a certain extent in a conflict between their fundamental ideological bent, which is Gifford Pinchot's belief that natural resources should be used wisely for the benefit of humans, and the symbolic message of the AWA. Pinchot's approach to conservation fits well with the ESA, legislation that essentially embodies the idea that humans can manage animal populations on an international scale. Pinchot's approach was not problematic before the AWA symbolically revalued the welfare of *individual* animals. Before that point, there seemed to be plenty of animals that one could, in theory, wisely manage for the benefit of humans. If one killed a few animals along the way to make space for the young, there were no philosophical conflicts. This perspective became a problem for zoos, however, as the animal protection and environmentalist movements swept the United States in the 1960s and 1970s and a spate of popular and scholarly studies demonstrated the extent to which humans had failed to manage nature's bounty wisely. Zoo directors were sympathetic to these conservationist concerns and incorporated them into their mission. Likewise, interest groups concerned with the welfare of all animals, rather than just the species, took issue with the idea that saving the group necessarily requires killing individual animals. These philosophical arguments were codified into national legislation with the AWA. For all of its limitations, the AWA gave symbolic weight to the idea that *all* animals deserve humane care, not just the ones that have the right kinds of genetic diversity. Zoo professionals are caught in these conflicting legislative regimes, between Pinchot's vision of managing natural resources and the values of the AWA. We should not be surprised that zoo professionals continue to struggle to reconcile these two conflicting legislative regimes and the different philosophical underpinnings of each one.

The final lesson of this story is that given the AZA's fundamental interest in protecting the welfare of exotic animals, the USDA was not the perfect fit for a zoo regulatory agency. Wild and exotic animals would have been much better off in a regulatory sense if Senator Mark Hatfield's bill had been enacted, because it would have created an agency whose primary constituency was zoos. In contrast, the history of zoos and regulatory politics since the 1970s is best characterized as an attempt to force the USDA, whose central constituency is farmers, to incorporate the job of overseeing zoos and the kinds of animals they hold. This is a common prescription for unhappiness on all sides as bureaucrats come under criticism for their lack of attention to the new group that they never wanted to regulate in the first place.[16] And those who are regulated feel that the agency that is regulating them really has other priorities. This helps us understand why the USDA's relationship to zoological institutions since the 1970s variously consisted of withdrawal, conflict, and ready acceptance of the zoo and aquarium community's suggestions for regulation revisions.[17]

Zoos will continue to regulate themselves, to a large extent because they have to make difficult scientific judgments.[18] Few people know about the problems that—for example—ectoparasites on elephants can cause, and thus regulations on these topics are easily determined by the regulated. Although it is fair to say that the USDA has deferred to zoos and aquariums on a host of matters affecting them, it is important to stress that this is what Congress intended. Congress wanted to regulate zoos and aquariums with the AWA amendments, but they also wanted to support these same institutions.

What is clear, however, is that Congress's intent will not prevent ongoing battles with and against animal protection groups. The AZA and animal protection groups will argue their respective positions before Congress, the courts, the bureaucracy, and, increasingly, state and local governments. The winners and losers in these battles will shape the future of American zoos and aquariums.

We can see the future of zoo politics in the post–Iraq War plight of the animals at the Baghdad Zoo. When American soldiers worked their way through the city's zoo, they found two empty cages bearing the words "Beagle" and "Pointer." The image of a small beagle, cowering in his cage as the city around him shook from the U.S. "shock and awe" bombing campaign encapsulates the twenty-first-century American zoo. In its cages, however elaborate and natural looking, live helpless animals who require constant care and attention. The U.S. military, much evolved from its Viet Nam days, assumed responsibility for the care of the Baghdad Zoo's captive creatures. Ultimately, such animals cannot escape the political world in which their keepers reside.[19] In a democracy, that political world includes many citizens who come to know their zoo animals as individuals, as animals analogous to the family beagle. These citizens are also cognizant of the fact that, when animals are housed in publicly owned institutions, the citizens do, in some sense, own them. Thus, it should not be surprising that at least some citizens feel grateful to those particular animals (but not necessarily to the whole species) for lifting their hearts a little bit on a Saturday afternoon. They would like to see these animals live as many years as possible, but if euthanasia decisions need to be made, or transfers to another kind of institution need to occur, they would like to be informed. Protecting animals and, by extension, the zoos in which they live, is a public mission, one that citizens largely support and that, at their best, zoos articulate and defend.

Notes

Introduction—The Political Revolution

1. *Blanche Guzzi v. New York Zoological Society,* 192 A.D. 263; 182 N.Y.S. 257 (1920) N.Y. App. Div. Lexis 7473.

2. *Animal Protection Institute of America v. Robert Mosbacher,* 799 F. Supp. 173; 1992 U.S. Dist. The plaintiffs also included the American Humane Association, Greenpeace, the Fund for Animals, and several organizations dedicated to whale or oceans conservation. Documents relating to the case used here include: Exhibit A "News from Shedd Aquarium, Fact Sheet" from Response of Petitioners Animal Protection Institute to Motion for Leave to Intervene by American Association of Zoological Parks and Aquariums; Motion for Summary Judgment by John Shedd Aquarium, F. 30; Response to Petitioners Animal Protection Institute, et al. to Motion for Leave to Intervene by American Association of Zoological Parks and Aquariums; Reply of AAZPA to Response of Animal Protection Institute, et al. to Motion for Leave to Intervene of AAZPA; Motion for Leave to Intervene by American Association of Zoological Parks and Aquariums; New York Zoological Society Brief of Amicus Curiae; Reply by John Shedd Aquarium, February 28, 1990; Petitioner's Response to the Court's Information Request of July 13, 1992; Opposition by John G. Shedd Aquarium to Petitioners' Motion for Summary Judgment.

3. "Aquarium Employees Face Charges From Cruel Capture of Belugas," *Animal Welfare Institute Quarterly* 39, no. 4 (Winter 1990/91): 7.

4. "Board of Directors Meeting," *AAZPA Newsletter* 29, no. 11 (November 1988): 5; "Annual Business and Awards Presentation Meeting," *Communiqué* (November 1991): 11.

5. Although there are two other large professional associations connected with zoos, the American Association of Zoo Keepers (AAZK) and the American Association of Zoo Veterinarians (AAZV), we focus on the AAZPA because it was the group that represented zoos in the political system.

6. Elizabeth Hanson, *Animal Attractions: Nature on Display in American Zoos* (Princeton, N.J.: Princeton University Press, 2002); Vernon Kisling, ed., *Zoo and Aquarium History: Ancient Animal Collections to Zoological Gardens* (Boca Raton, Fla.: CRC Press, 2001).

7. Animal Welfare Institute (www.awionline.org, accessed November 17, 2004). We chose to use the term "animal protection groups" when we refer to both animal welfare and rights groups that worked against zoos. When we are discussing a particular animal group that is clearly within one camp or the other, we refer to it as that type of group. For example, we refer to the Humane Society of the United States as an animal welfare group. There is no systematic study of all animal protection groups in the United States, so we had to determine whether they were welfare- or rights-based groups. A few of them have been described in Margaret C. Jasper's *Animal Rights Law* (Dobbs Ferry, N.Y.: Oceana Publications, 1997), but many of them have received little analysis. When a group was not already described as welfare- or rights-based we looked at their promotional materials. We recognize, however, that these are rough typologies.

8. Vernon N. Kisling, Jr., "The Origin and Development of American Zoological Parks to 1899" in R. J. Hoage and Willam A. Deiss, eds., *New Worlds, New Animals: From Menagerie to Zoological Park in the Nineteenth Century* (Baltimore: Johns Hopkins University Press, 1996); William Bridges, *Gathering of Animals: An Unconventional History of the New York Zoological Society* (New York: Harper and Row, 1974); Jeffrey Nugent Hyson, "Urban Jungles: Zoos and American Society," Ph.D. dissertation, Cornell University, 1999; Jeffrey Stott, "The Historical Origins of the Zoological Park in American Thought," *Environmental Review* 5 (Fall 1981): 52–65.

9. E. E. Schattschneider, *The Semisovereign People: A Realist's View of Democracy in America* (New York: Holt, Rinehart, and Winston, 1960), 35.

10. Biographies are available at the AZA's Web site, www.aza.org.

11. Earl Latham, "The Group Basis of Politics: Notes for A Theory," *American Political Science Review* 46, no. 2 (June 1952): 376–97.

1—Opening Moves

1. Fairfield Osborn, "Another Noah's Ark—For New York," *New York Times* (October 27, 1963), ProQuest Historical Newspapers database; AAZPA Executive Board Meeting Minutes (October 8, 1966): 3, Box 9, AZA Record 99-102.

2. W. Douglas Costain and James P. Lester. "The Evolution of Environmentalism," in James P. Lester, ed., *Environmental Politics and Policy: Theories and Evidence* (Durham, N.C.: Duke University Press, 1995), 15–38.

3. Marc Bekoff and Jan Nystrom, "The Other Side of Silence: Rachel Carson's View of Animals," *Zygon* 39, no. 4 (December 2004): 861–83, esp. 863 and 864.

4. AZA Executive Board Meeting Minutes (October 8, 1966): 5, Box 9. AZA Record 99-102; Theodore H. Reed, "Report of the AAZPA Conservation Committee and Subcommittee on Endangered Species," AAZPA Executive Board Meeting Minutes (December 1 and 2, 1967): 33–34, Box 9, AZA Record 99-102.

5. AZA Executive Board Meeting Minutes (October 8, 1966): 4, Box 9, AZA Record 99-102.

6. AZA Executive Board Meeting Minutes (February 19, 1966): 12–13, Box 9, AZA Record 99-102; AZA Executive Board Meeting Minutes (October 8, 1966): 4, Box 9, AZA Record 99-102; Fred Zeehandelaar, "Report to the Executive Board AAZPA" (March 14, 1967): A-16, Box 9, AZA Record 11-102. AAZPA Executive Board Meeting Minutes (February 14–15, 1970) Box 9, AZA Record 99-102.

7. AZA Executive Board Meeting Minutes (October 8, 1966): 3, Box 9, AZA Record 99-102.

8. Theodore H. Reed, "Report of the AAZPA Conservation Committee and Subcommittee on Endangered Species," AAZPA Executive Board Meeting Minutes (December 1 and 2, 1967): 32, Box 9, AZA Record 99-102; Executive Board Meeting Minutes (February 19, 1966): 9–10, Box 9, AZA Record 99-102; Executive Board Meeting Minutes (October 8, 1966): 7 and 11, Box 9, AZA Record 99-102.

9. "Resolution Protecting Certain Endangered Species," AZA Executive Board Meeting Minutes, (March 11 and 14, 1967): A-1, Box 9, AZA Record 99-102.

10. "Resolution Protecting Certain Endangered Species," AZA Executive Board Meeting Minutes, (March 11 and 14, 1967): A-2, Box 9, AZA Record 99-102.

11. AAZPA Executive Board Meeting Minutes (December 1 and 2, 1967): 4, 32–33, Box 9, AZA Record 99-102.

12. Brian Czech and Paul R. Krausman, *The Endangered Species Act: History, Conservation, Biology, and Public Policy* (Baltimore: Johns Hopkins University Press, 2001).

13. Executive Board Meeting Minutes (February 19, 1966): 9, Box 9, AZA Record 99-102; Executive Board Meeting Minutes (March 11–12, 1967): 6, Box 9, AZA Record 99-102; John Perry (assistant director, National Zoo) to Carol Chase, September 4, 1968, Box 36, "Import/Export Committee 1960–1969," AZA Archives, Record 96-024.

14. Frederik J. Zeehandelaar and Paul Sarnoff, *Zeebongo: The Wacky Wild Animal Business* (Englewood Cliffs, N.J.: Prentice-Hall, 1971), 169; "Zeehandelaar Indicted," *Animal Welfare Institute Information Report* 21, no. 4 (1972): 4 (hereafter *AWIIR*).

15. Fred J. Zeehandelaar, "Report to the Executive Board AAAZPA," Meeting Minutes (May 11–12, 1968): 3, Box 9, AZA Record 99-102.

16. Theodore H. Reed, "Report of the AAZPA Conservation Committee and Subcommittee on Endangered Species," AAZPA Executive Board Meeting Minutes (December 1 and 2, 1967): 32, Box 9, AZA Record 99-102.

17. Executive Board Meeting Minutes (March 11–12, 1967): 6, Box 9, AZA Record 99-102; Fred Zeehandelaar, "Report to the Executive Board AAZPA" (May 11–12, 1968): 3, Box 9, AZA Record 99-102; Executive Board Meeting Minutes (May 11–12, 1968), Box 9, AZA Record 99-102; John Perry to Carol Chase, December 4, 1968, Box 36, "Import/Export Committee 1960–1969," AZA Archives, Record 96-024.

18. Czech and Krausman, *The Endangered Species Act,* 22.

19. Statement of Theodore Reed (director, National Zoological Park), House Subcommittee on Fisheries and Wildlife Conservation, *Fish and Wildlife Legislation Part I: Hearings before the House Subcommittee on Fisheries and Wildlife Conservation,* 90th Cong., 1st sess., October 4, 1967, 41–60.

20. AAZPA Annual Meeting of Board of Directors Minutes (September 18–19, 1971): 23, Box 9, AZA Record 99-102.

21. F. Wayne King to Gary K. Clarke (March 30, 1972): 1–2, Box 9, 96-024.

22. F. Wayne King, "Report of the AAZPA Wildlife Conservation Committee to the AAZPA Board of Directors" (March 1974): addendum, AAZPA Board of Directors Meeting Minutes (March 15, 1974), Box 9, AZA Record 99-102.

23. Executive Board Meeting Minutes (October 8, 1966) AZA, Box 9, Record 99-102.

24. Frank M. Thompson to members of the AAZPA Importation Committee, December 4, 1965, "Import/Export Committee 1960–1969," Box 36, AZA Record 96-024.

25. Bernard Unti, *Protecting All Animals: A Fifty-Year History of the Humane Society of the United States* (Washington, D.C.: Humane Society Press, 2004), 119–20; Carolyn Etter and Don Etter, *The Denver Zoo: A Centennial History* (Boulder, Colo.: Rinehart, 1995), 130.

26. Executive Board Meeting Minutes (October 8, 1966), Box 9, AZA Record 99-102; Robert Wagner, 1996 Proceedings, 393–94, Box 4, AZA Record 99-102; William G. Conway to F. J. Mulhern, January 26, 1967, Conway to Mulhern, March 3, 1967, Box 36, "Import/Export Committee 1960–1969," AZA Record 96-204.

27. Executive Board Meeting Minutes (March 11–12, 1967): A-14, A-16, Box 9, AZA Record 99-102. The Kiang and Tiang are two different kinds of antelopes.

28. Executive Board Meeting Minutes (May 12, 1968), Box 9, AZA Record 99-102.

29. Board of Directors Meeting Minutes (September 18–19, 1971): 3–4, Box 9, AZA Record 99-102.

30. "Schweitzer Medal Presented to Dr. Mulhern and Dr. Jones by Senator Holland," *AWIIR* 16, no. 4 (1967): 1–4.

31. Board of Directors Meeting Minutes (September 18–19, 1971): 3–4, Box 9, AZA Archives, Record 99-102.

32. Zeehandelaar and Sarnoff, *Zeebongo,* 158.

33. House Subcommittee on Livestock and Grains of the Committee on Agriculture, *Care of Animals Used for Research, Experimentation, Exhibition, or Held for Sale as Pets,* 91st Cong., 2nd sess., 1970 (Washington, D.C.: Government Printing Office, 1970), 39. Statement of Chris Sullivan, "Again, Zoo Situation."

34. House Subcommittee on Livestock and Grains, *Care of Animals,* Serial DD. For the Statement of Cecile B. O'Marr, Field Representative, Defenders of Wildlife, Washington, D.C., see 92–93.

35. Jeffrey Smith, "Oh, Let Them Out!" *Defenders of Wildlife News* (1968): 395–99; Michael Frome, "Roadside Zoos," *Defenders of Wildlife News* (1964): 13–14.

36. Unti, *Protecting All Animals,* 1–40.

37. "Congressman Whitehurst Introduces Five Bills for Animal Welfare," *AWIIR* 18, no. 4 (1969): 1–3.

38. Interview with G. William Whitehurst, March 26, 2004; interview with Jane Whitehurst, March 29, 2004.

39. "Hearings Held on Whitehurst Bill to Broaden Laboratory Animal Welfare Act," *AWIIR* 19, no. 2 (1970): 1–2.

40. Ronald T. Reuther to Bob Truett, May 17, 1970, Reorganization Committee AAZPA-NRPA 1970–71, Box 36, AZA Record 96-024.

41. Lester Fisher to Gunter Voss, November 24, 1970, Reorganization Committee AAZPA-NRPA 1970–71, Box 36, AZA Record 96-024.

42. Don Bohnet to Ira Hutchison, July 23, 1970, Legislation Committee 1971, Box 36, AZA Record 96-024.

43. House Subcommittee on Livestock and Grains, *Care of Animals,* 34, 38–39; statement by Christine Stevens, President, Animal Welfare Institute, and Secretary for Animal Protective Legislation, Washington D.C.

44. "Cranston-Magnuson-Spong Bill for Warm-Blooded Animals," *AWIIR* 19, no. 3 (1970): 1–2.

45. P. W. Ogilvie to Gunter Voss, December 31, 1970, Legislation Committee 1971, Box 36, AZA Record 96-024; Robert M. Artz to Officers and Board Members, December 11, 1970, Legislation Committee 1971, Box 36, AZA Record 96-024.

46. Linda Koebner, *Zoo: The Evolution of Wildlife Conservation Centers* (New York: Forge, 1994), 169; *Animal Welfare Act,* sections 7, 8, 10, 13.

47. House Subcommittee on Livestock and Grains, *Care of Animals,* 106 (section 13 of the act).

48. House Subcommittee on Livestock and Grains, *Care of Animals*, 44.

49. John Perry to Philip Ogilvie, January 4, 1971, Reorganization Committee AAZPA-NRPA 1970–71, Box 36, AZA Record 96-024: Robert M. Artz to Officers and Board Members, December 11, 1970, Legislation Committee 1971, Box 36, AZA Record 96-024.

50. AAZPA, AWA and USDA, Box 13, AZA Record 96-024; Robert Wagner, "Executive Board Minutes, Mid-Year Activity Report, Legislative Committee," (March 15, 1974), 2, Box 9, AZA Record 99-102.

51. "Zoo Federation Study," (1971), Zoo Federation, Box 28, AZA Record 96-024; Saul Kitchener to Gark Clarke, April 29, 1970, Reorganization Committee AAZPA-NRPA 1970–71, Box 36, AZA Record 96-024; Ronald Reuther to Willard Brown, March 23, 1971, Reorganization Committee AAZPA-NRPA 1970–71, Box 36, AZA Record 96-024; W. G. Conway and T. H. Reed to AAZPA, August 25, 1970, History, Box 18, AZA Record 96-024.

52. Board of Directors Meeting Minutes (February 14–15, 1970), Box 9, AZA Record 99-102; Board of Directors Meeting Minutes (March 7–8, 1971), Box 9, AZA Record 99-102; Clyde Hill, "Report, Editorial Committee" (September 1971), Box 3, AZA Archives, Record 99-102; "NRPA Legislative and Public Affairs Action Report—March, 1971," Legislation Committee 1971, Box 36, AZA Record 96-024.

53. Robert Wagner, "The Independence of AAZPA," *Proceedings of the 1996 Annual Conference,* 393–96, Box 3, AZA Record 99-102.

54. Gunter Voss, "State Statutes," Board of Directors Meeting Minutes (1969) 23, AZA Archives, Box 9, Smithsonian Institution; P. W. Ogilvie to Gunter Voss, December 31, 1970, Legislation Committee 1971, Box 36, AZA Record 96-024.

55. Robert Wagner, e-mail communication with the authors, April 6, 2004.

56. Articles of Incorporation (December 14, 1971): 1–3, AZA Record 99-102.

57. Varying membership dues is a common interest group tactic. Jack L. Walker, "The Origin and Maintenance of Interest Groups in America," *American Political Science Review* 77, no. 2 (June 1983): 390–406.

58. Articles of Incorporation (December 14, 1971): 1–3, AZA Record 99-102.

59. William Conway to Gunter Voss, February 17, 1971, Reorganization Committee AAZPA-NRPA 1970–71, Box 36, AZA Record 96-024.

2—On the Defensive

1. Fred Zeehandelaar, "Report," AAZPA Board Meeting Minutes (April 15–16, 1972): 1–2, Box 9, AZA Record 99-102.; D. F. McMichael to F. J. Zeehandelaar (April 6, 1972), and F. J. Zeehandelaar to D. F. McMichael, (April 12, 1972), both included in the AAZPA Board Meeting Minutes, April 15–16, Box 9, AZA Record 99-102.

2. Christopher Bosso, *Pesticides and Politics: The Life Cycle of a Public Issue* (Pittsburgh: University of Pittsburgh Press, 1987).

3. "Subject List of Federal Bills Affecting Zoos" [1973], Box 2, AAZPA Legislative Committee, 1973–1974, AZA Record 96-024.

4. Ronald Reuther to Bob Truett, May 17, 1970, Reorganization Committee AAZPA-NRPA 1970–71, Box 36, AZA Record 96-024.

5. Warren Iliff, "Zoo Federation Study" (1971), 12, 33 and 37, Zoo Federation, Box 28, AZA Record 96-024; George Rabb to Colleagues, June 25, 1970, History, Box 18, AZA Record 96-024; Lester Fisher to Gunter Voss, November 24, 1970, Reorganization Committee AAZPA-NRPA 1970–71, Box 36, AZA Record 96-024.

6. Fred Zeehandelaar to Lester Fisher, June 29, 1973, AAZPA Legislative Committee 1973–1974, Box 2, AZA Record 96-024; AAZPA Board of Directors Mid-Year Meeting Minutes (March 29–30, 1973), Box 9, AZA Record 99-102.

7. "Two Fish and Wildlife Service Officials Receive New Appointments" (October 5, 1954) and "Interior Issues 'Preview' of National Fisheries Center" (December 9, 1963), U.S. Fish and Wildlife (FWS) press releases (news.fws.gov/historic, accessed May 18, 2005).

8. William Hagen to William Braker, May 7, 1973, AAZPA Legislative Committee 1973–1974, Box 2, AZA Record 96-024.

9. Gary Clarke to Richard Denney, June 9, 1972; Richard Denney, "Supportive Data on Some United States Zoos for Legislative Assistance," September 30, 1972; Denney to AAZPA Officials, November 8, 1972; R. L. Blakely to Rutherford Phillips (AHA), September 16, 1974; AHA to Blakely (mailgram), October 1, 1974; all in American Humane Association, Box 12, AZA Record 96-024.

10. Ronald Reuther to Godfrey Rockefeller (WWF), February 27, 1975; Rockefeller to Reuther, March 14, 1975; both in World Wildlife Fund 1980–85, Box 28, AZA Record 96-024. "Symposium on Endangered and Threatened Species in North America," 93rd Cong., 2nd sess., *Congressional Record* 120, no. 60 (May 1, 1974): 2681–82.

11. Quotation comes from photo caption (np) in Lewis Regenstein (Fund for Animals executive vice president), *The Politics of Extinction: The Shocking Story of the World's Endangered Wildlife* (New York: MacMillan, 1975).

12. Regenstein, *Politics,* 100–108.

13. "Hearings on Bills to Protect Sea Mammals," *AWIIR* 20, no. 3 (1971); "Marine Mammals Legislation," *AWIIR* 20, no. 4 (1971): 1–4.

14. Statement by John Prescott, Senate Subcommittee on Oceans and Atmosphere, *Ocean Mammal Protection: Hearings before the Senate Subcommittee on Oceans and Atmosphere,* 92nd Cong., 2nd sess., February 15, 16, and 23, and March 7, 1972, 547–55.

15. "Marine Mammal Protection Act Report," *AWIIR* 22, no. 3 (1973): 3; Regenstein, *Politics,* 113–14.

16. Regenstein, *Politics,* 137–50.

17. 93rd Cong., 1st sess., *Congressional Record* 119 (July 24, 1973), pt. 20: 25668–69, 25675; Joe Mann, "Making Sense of the Endangered Species Act: A Human-Centered Justification," *New York Environmental Law Journal* 7 (1999): 246–305.

18. Nathanial P. Reed, speech before the HSUS, October 19, 1973, History, Box 17, AZA Record 96-024.

19. Nathaniel Reed to Gary Clarke, August 15, 1974; Gary Clarke to Nathaniel Reed, September 9, 1974; and other letters in History, Box 17, AZA Record 96-024.

20. Regenstein, *Politics;* "Lawsuits are the Only Alternative Left," *Fish and Wildlife Service News* (February 1975): 7–8.

21. "Saving Wildlife by Enforcing the Law," *AWIIR* 23, no. 2 (1974): 1–2; "Interior Proposes Major Restrictions on Wildlife Imports," *AWIIR* 23, no. 1 (1974): 1; "Proposed Regulations May Hurt Primate Imports for Scientific Use," *National Society for Medical Research Bulletin* 25, no. 2 (February 1974): 1–2. The Proposed Importation Regulations on Injurious Wildlife were published in the *Federal Register* 38, no. 244 (December 20, 1973).

22. L. E. Fisher to Peg Dankworth (AAZPA executive director), July 3, 1973, Box 2, AAZPA Legislative Committee, 1973–1974, AZA Record 96-024; Robert Wagner, "Mid-Year Activity Report, Legislative Committee" (March 15, 1974): 1–7, AAZPA

Board of Directors Meeting Minutes (March 15, 1974), Box 9, AZA Record 99-102.

23. "AAZPA Position and Comments: Proposed Importation Regulations on Injurious Wildlife" (1974), Box 9, AZA Record 99-102.

24. "Saving Wildlife by Enforcing the Law," *AWIIR* 23, no. 2 (1974): 1–2.

25. AAZPA Board of Directors Meeting Minutes (September 27–29, 1974): 7–8, AAZPA Board of Directors Meeting Minutes (March 15, 1974), Box 9, AZA Record 99-102; Robert Wagner, "Annual Activity Report Legislative Committee" (September 28–29, 1974): 2–3; AAZPA Board of Directors Meeting Minutes (September 27–29, 1974); AAZPA Board of Directors Meeting Minutes (March 15, 1974), Box 9, AZA Record 99-102; Wagner, "Legislative Committee Report" (October 3, 1974): 1–2; AAZPA Board of Directors Meeting Minutes (September 27–29, 1974); AAZPA Board of Directors Meeting Minutes (March 15, 1974), Box 9, AZA Record 99-102.

26. William Braker, "Legislative Chairman's Report," *AAZPA Newsletter* 16, no. 7 (July 1975): 3–4; "Interior Revives Injurious Wildlife Plan," *AAZPA Newsletter* 17, no. 9 (September 1976): 6.

27. AAZPA Board of Directors Meeting Minutes (October 6–7 and 10, 1973), Box 9, AZA Record 99-102.

28. Bob Truett, "USDA and Zoos: An Editorial Opinion," *Animal Tracks* 5, no. 2 (March 15, 1972); AAZPA Board of Directors Minutes (1971–72), Box 9, AZA Record 99-102; Frank Powell to Margaret Dankworth, August 30, 1974, ZooAct, Inc., Box 36; "America's Zoos" (ZooAct press release, 1975), AAZPA Legislative Committee 1976, Box 2, AZA Record 96-024.

29. Lester Fisher to Peg Dankworth, July 3, 1973, AAZPA Legislative Committee 1973–1974, Box 2, AZA Record 96-024.

30. Robert Wagner, "Annual Activity Report Legislative Committee" (September 28–29, 1974): 1, Box 9, AZA Record 99-102; Senator Mark Hatfield of Oregon introducing S 2042, 93rd Cong., 1st sess., *Congressional Record* 119 (June 20, 1973), pt. 16:20441.

31. Mark O. Hatfield, *Against the Grain: Reflections of a Rebel Republican* (Ashland, Ore.: Whitecloud Press) 214; "'Just Imagine,'" *Animal Welfare Institute Quarterly* 30, no. 4 (Winter 1982): 6–7.

32. Senate Subcommittee on the Smithsonian Institution of the Committee on Rules and Administration, *Federal Assistance for Zoos and Aquariums: Hearing before the Subcommittee on the Smithsonian Institution of the Committee on Rules and Administration,* S 2042 and S 2774, 93rd Cong., 2nd sess., 1974, 5–8.

33. Ronald Reuther to AAZPA Legislative Committee, May 30, 1973, AAZPA Legislative Committee 1973–1974, Box 2, AZA Record 96-024; Robert Wagner, "Legislative Chairman's Report," (September 1974): 4, Box 9, AZA Record 99-102.

34. Robert Wagner, "Executive Board Minutes, Mid-Year Activity Report, Legislative Committee," (March 15, 1974): 3, Box 9, AZA Record 99-102; Senate, *Federal Assistance for Zoos and Aquariums,* 46, 73.

35. Senate, *Federal Assistance for Zoos and Aquariums,* 5–6, 39, 71, 82.

36. Senate, *Federal Assistance for Zoos and Aquariums,* 61, 66, 82.

37. Senate, *Federal Assistance for Zoos and Aquariums,* 36.

38. Bernard Fensterwald, "Time to Phase Out Zoos," *Washington Post* (February 2, 1974).

39. Senate, *Federal Assistance for Zoos and Aquariums,* 37–39, 66; Margaret Dankworth to Lester Fisher, May 8, 1973, AAZPA Legislative Committee 1973–1974, Box 2, AZA Record 96-024; G. William Whitehurst to Lester Fisher, October 16, 1973, United Action for Animals, Box 27, AZA Record 96-024.

40. Senate, *Federal Assistance for Zoos and Aquariums,* 39–41, 56.

41. House, Representative G. William Whitehurst of Virginia speaking on A Retreat for Rare Animals, 93rd Cong., 2nd sess., *Congressional Record* 120 (February 20, 1974), pt. 3:3641.

42. House, Representative G. William Whitehurst of Virginia speaking on Humane Care in Zoos, 94th Cong., 1st sess., *Congressional Record* 121 (March 10, 1975), pt. 5:5935.

43. "New Zoo Control Bill Reintroduced," *AAZPA Newsletter* 19, no. 5 (May 1978): 5; "National Zoological Foundation," *AAZPA Newsletter* 20, no. 3 (March 1979): 6.

44. House, Representative John Dingell of Michigan introducing HR 70, 94th Cong., 1st sess., *Congressional Record* 121 (January 14, 1975), pt. 1:146.

45. Prepared Statement for the authors by Congressman John Dingell, April 29, 2004.

46. Robert Wagner, "Legislative Chairman's Report," (September 1974): 4, Box 9, AZA Record 99-102.

47. House, Representative John Dingell of Michigan introducing HR 70, 94th Cong., 1st sess., *Congressional Record* 121 (January 14, 1975), pt. 1:146; William Hagen, "Legislation," *AAZPA Newsletter* 16, no. 2 (February 1975): 6; Gordon Hubbell, "Annual Report Special Committee to Work with the USDA to Determine Minimum Cage Sizes for Exhibiting Animals" (September 27, 28, and 29, 1974), Box 9, AZA Record 99-102; John Mehrtens, "Can Zoos Survive?" (article clipping, 1974), John Mehrtens, Box 21, AZA Record 96-024; William Braker, "Big Brother is Watching the Animals," February 21, 1975, History, Box 17, AZA Record 96-024.

48. Board of Directors Meeting Minutes (September 27–29, 1974): 4 Box 9, AZA Record 99-102; "Legislation," *AAZPA Newsletter* 16, no. 2 (February 1975): 6; see also March and May *Newsletters.*

49. Miscellaneous brochures, ZooAct, Inc., Box 36; "America's Zoos" (ZooAct press release, 1975), AAZPA Legislative Committee 1976, Box 2; John Mehrtens, "Can Zoos Survive?" (article clipping, 1974), John Mehrtens, Box 21, AZA Record 96-024.

50. Senate, Senator Ernest Hollings of South Carolina speaking on Zoos are for Animals, 93rd Cong., 2nd sess., *Congressional Record* 120 (September 25, 1974), pt. 24: 32560; House, Representative Floyd Spence of South Carolina speaking on Bureaucratic Case for Zoos?, 93rd Cong., 2nd sess., *Congressional Record* 120 (October 2, 1974), pt. 25:33686; Senate, Senator Strom Thurmond of South Carolina speaking on Zoo Management, 93rd Cong., 2nd sess., *Congressional Record* 120 (October 3, 1974), pt. 25:33799.

51. Robert Wagner, "Legislative Committee Report," October 3, 1974, Box 9, AZA Record 99-102 (emphasis added).

52. William Braker (AAZPA president) to John Prescott (director, New England Aquarium), March 28, 1974, Box 9, AZA Record 99-102; David B. Truman, *The Governmental Process: Political Interests and Public Opinion* (New York: Alfred A. Knopf, 1951), 156–87.

53. Braker to Prescott, March 28, 1974, Box 9, AZA Record 99-102; John Prescott (chairman, Special Committee for Reorganization) to William Braker, September 20, 1974, Box 9, AZA Record 99-102.

54. "Getting Down to a Few Brass Tacks: An Open Letter to AAZPA Members," February 14, 1975, John Mehrtens, Box 21, AZA Record 96-024.

55. Robert Wagner, "Legislative Committee Report," October 3, 1974, Box 9, AZA Record 99-102.

3—A Stronger Zoo Community

1. R. L. Blakely, "A Letter to Zoo Societies," *AAZPA Newsletter* 16, no. 2 (February 1975): 2, and "From the President's Desk," *AAZPA Newsletter* 16, no. 3 (March 1975): 1.

2. Blakely, "From the President's Desk"; R. L. Blakely, "State of the Association," *AAZPA Newsletter* 16, no. 5 (May 1975): 2, and "An Open Letter to the HSUS," *AAZPA Newsletter* 16, no. 6 (June 1975): 4.

3. Frank Powell to Margaret Dankworth, August 30, 1974; Robert Wagner to Clyde Hill, 9 May 1975; Ronald Blakely to Warren Iliff, July 9, 1975, ZooAct, Inc., Box 36, AZA Record 96-024; *AAZPA Newsletter* 16, no. 12 (December 1975): 2.

4. "Statement of Accountability," *AAZPA Newsletter* 17, no. 4 (April 1976): 2.

5. "Importation of Wild Animals and Birds," *AWIIR* 12, no. 3 (1963).

6. Philip Ogilvie, "Making the Most of Our Zoos," *Parks and Recreation* 4 (January 1969): 32–34, 50; Gary Clarke, "European Zoo Marathon," *Parks and Recreation* 4 (October 1969): 41–42; Ronald Reuther, "A New Zoo View," *Parks and Recreation* 6 (September 1971): 17–19, 54–56; Cleveland Amory, *Man Kind? Our Incredible War on Wildlife* (New York: Harper and Row, 1974), 323–25.

7. Leonard J. Goss. "Voluntary Registration Plan, Minutes, Executive Board" (October 8, 1966): 27, Box 9, AZA Record 99-102; Robert Wagner, "Annual Activity Report Legislative Committee" (September 28 and 29, 1974): 6 Box 9, AZA Record 99-102.

8. "More Institutions Accredited by AAZPA Commission," *AAZPA Newsletter* 17, no. 8 (August 1976): 1.

9. "Commission and Advisor Reports," *AAZPA Newsletter* 18, no. 6 (June 1977): 10.

10. "Tulsa Opposes H.R. 6631," *AAZPA Newsletter* 16, no. 8 (August 1975): 5; "Something to Consider," *AAZPA Newsletter* 16, no. 5 (May 1975): 5.

11. Robert Wagner, "AAZPA Executive Director Report," *AAZPA Newsletter* 17, no. 7 (July 1976): 8; William Hagen, "Federal Regulations," *AAZPA Newsletter* 16, no. 8 (August 1975): 4; Paul Chaffee to Robert Wagner, October 17, 1977, Political Action Workshop, Box 23, AZA Record 96-024.

12. Donald E. Veraska, "Zoos Deaths May Spur U.S. Suit," *Hackensack Record* (March 2, 1975), reprinted in *AWIIR* 24, no. 2 (May 1975): 4.

13. 93rd Cong., 1st sess., *Congressional Record* 119 (July 24, 1973), pt. 20:25668–69, 25675–76, 25679, 25693.

14. AAZPA Board of Directors Meeting Minutes, October 6–7, 10, 1973, Box 9, AZA Record 99-102.

15. Robert Wagner, "Brief Legislative Report," *AAZPA Newsletter* 16, no. 5 (May 1975): 4; William Braker, "Legislation," *AAZPA Newsletter* 16, no. 7 (July 1975): 3; William Braker, "Legislation," *AAZPA Newsletter* 16, no. 11 (November 1975): 6; "AAPZA Response to USDI/FWS Proposal 'Determination of Captive, Self-Sustaining Populations,'" *AAZPA Newsletter* 17, no. 8 (August 1976): 3–4.

16. Robert Wagner to Board of Directors, October 6, 1975, Endangered Species Oversight Hearings, Box 16, AZA Record 96-024; *AAZPA Newsletter* 16, no. 11 (November 1975): 1, 5–6; U.S. House, "Endangered Species Oversight Hearings before the Subcommittee on Fisheries and Wildlife Conservation and the Environment of the Committee on Merchant Marine and Fisheries," 94th Cong., 1st sess. (October 1, 2, and 6, 1975): 131–35, 143, 215.

17. "Sixteen Species Proposed as Captive Self-Sustaining," *AAZPA Newsletter* 17, no. 4 (April 1976): 5; Robert Wagner, "USDI/FWS Publishes Final Rulemaking to Establish Captive, Self-Sustaining Populations," *AAZPA Newsletter* 18, no. 7 (July 1977): 3.

18. Statement of Robert Wagner, Senate Subcommittee on Resource Protection, *Endangered Species Act Oversight Hearings before the Subcommittee on Resource Protection of the Senate Committee on Environment and Public Works,* 95th Cong., 1st sess., July 20, 21, 22, and 28, 1977, 552–56.

19. Robert Wagner, "AAZPA Executive Director's Report," *AAZPA Newsletter* 19, no. 7 (July 1978): 3; "News from Washington," *AAZPA Newsletter* 19, no. 11 (November 1978): 22; Edward Maruska and Robert Wagner to Cecil D. Andrus (June 1, 1979), *AAZPA Newsletter* 20, no. 7 (July 1979): 4.

20. Tom Foose, "Conservation Coordinator's Report," *AAZPA Newsletter* 22, no. 4 (April 1981): 8.

21. "Information on AAZPA's Species Survival Plan," *AAZPA Newsletter* 21, no. 12 (December 1980): 4–5.

22. Dennis Meritt, "Midwest Fish and Wildlife Conference," December 1978, Wildlife Conservation 1979, Box 38, AZA Record 96-024.

23. Meritt, "Midwest Fish and Wildlife Conference."

24. Zoological Environmental Conference [I], Box 29, AZA Record 96-024.

25. See Bonnie Burgess, *Fate of the Wild: The Endangered Species Act and the Future of Biodiversity* (Athens: University of Georgia Press, 2001).

26. G. Steele, "News from Washington," *AAZPA Newsletter* 20, no. 10 (October 1979): 11.

27. E. Wolfe, "Report Released Revealing American Attitudes about Wildlife," *AAZPA Newsletter* 21, no. 1 (January 1980): 4.

28. "Reauthorization Hearings and Proposed Amendment to the Endangered Species Act," *AAZPA Newsletter* 19, no. 5 (May 1978): 4; "A Report on the Fifth Zoological/Environmental Conference," *AAZPA Newsletter* 19, no. 6 (June 1978): 1; Robert Wagner, "AAZPA Executive Director Report," *AAZPA Newsletter* 19, no. 8 (August 1978): 3; Senator John C. Culver to J Kevin Bowler, June 13, 1978, Legislative Committee 1971–80, Box 2, AZA Record 96-024. Wagner testified before the House Fisheries and Wildlife Conservation and the Environment Subcommittee on May 26, 1978.

29. "Zoological/Environmental Conference Held September 1, 1976," *AAZPA Newsletter* 17, no. 10 (October 1976): 3–4; Robert Wagner, "Second AAZPA/Environmental Conference," *AAZPA Newsletter* 18, no. 2 (February 1977): 1; Paul Chafee, "Cooperation Method Suggested," *AAZPA Newsletter* 18, no. 4 (April 1977): 12; Vernon Kisling, "Wildlife Society Annual Meeting Held," *AAZPA Newsletter* 18, no. 6 (June 1977): 22; George Steele, "Washington Legislative and Regulatory Zoological Highlights—1979," (September 29, 1979), Steele and Utz 1979, Box 38, AZA Record 96-024; E. Wolfe, "News from Washington," *AAZPA Newsletter* 20, no. 9 (September 1979): 10.

30. G. Steele, "Controversy Rages over Breaux's Proposed Amendment to Endangered Species Act," *AAZPA Newsletter* 20, no. 11 (November 1979): 5; E. Wolfe, "House and Senate Approve Appropriations for Endangered Species Act," *AAZPA Newsletter* 20, no. 12 (December 1979): 18. Kristin Vehrs, "News From Washington," *AAZPA Newsletter* 23, no. 4 (April 1982): 7.

31. 90th Cong., 2nd sess., *Congressional Record* 114 (August 1, 1968), pt. 19:24768; Fay Brisk, "Animals and Airlines," in Emily Leavitt, ed., *Animals and Their Legal Rights: A Survey of American Laws from 1641 to 1990* (Animal Welfare Institute: New York, 1990), 106–12.

32. Peter Batten, *Living Trophies* (New York: Thomas Y. Crowell, 1976), 19–20; Senate Hearing before the Subcommittee on the Environment of the Committee on Commerce, *Animal Welfare Improvement Act of 1975,* 94th Cong., 1st sess. (1975): 112;

House Subcommittee on Livestock and Grains of the Committee on Agriculture, *Care of Animals Used for Research, Experimentation, Exhibition, or Held for Sale as Pets,* 91st Cong., 2nd sess. (Washington, D.C.: Government Printing Office, 1970): 74.

33. "Recent Congressional Action on Protection of Animals," *AWIIR* 23, no. 1 (1974): 6.

34. Senate Subcommittee on the Environment of the Committee on Commerce, *Animal Welfare Improvement Act of 1975: Hearing Before the Subcommittee on the Environment of the Committee on Commerce, United States Senate,* S 1491, S 2070, S 2430, 94th Cong., 1st sess. (1975): 1.

35. Senate Subcommittee on the Environment of the Committee on Commerce, 94th Cong., 1st sess., *Animal Welfare Improvement Act of 1975* (1975): 112; Robert Wagner to Daniel Michalowski, October 22, 1974, AAZPA Legislative Committee 1971–1980, Box 2, AZA Record 96-024.

36. Senate, *Animal Welfare Improvement Act of 1975,* 35, 43, 51–52.

37. Richard P. Skully, Director, Flight Standards Service, Federal Aviation Administration, Senate, *Animal Welfare Improvement Act of 1975,* 48.

38. Senate, *Animal Welfare Improvement Act of 1975,* 101–3.

39. Senate, *Animal Welfare Improvement Act of 1975,* 114.

40. Senate, *Animal Welfare Improvement Act of 1975,* 106–7.

41. "Air Shipment Problems," *AAZPA Newsletter* 17, no. 1 (January 1976): 4; "CAB to Investigate Complaint," *AAZPA Newsletter* 17, no. 6 (June 1976): 4.

42. "C.A.B. Hearings," *AAZPA Newsletter* 16, no. 11 (November 1975): 8; "Further Notice on C.A.B. Hearings," *AAZPA Newsletter* 17, no. 1 (January 1976): 4.

43. "Animal Welfare Act Amendments Progress," *AWIIR* 25, no. 1 (1976): 3–4.

44. F. J. Zeehandelaar to Lester Fisher (President AAZPA), June 29, 1973, Box 2, AAZPA Legislative Committee, 1973–1974, Accession 96-024, AZA Archives. Robert Wagner, "Reflections upon 1976," *AAZPA Newsletter* 18, no. 1 (January 1977): 18.

45. Mark Rich, "AAZPA/Marine Mammal Commission: Are We Missing the Boat?" *AAZPA Newsletter* 16, no. 4 (April 1975): 2.

46. Marine Mammal Protection Act Oversight Hearings—1975, Box 2, AZA Record 96-024.

47. Don Bonker, "Protecting the Killer Whale," 94th Cong., 2nd sess., *Congressional Record* 122 (March 11, 1976), pt. 5.

48. U.S. House of Representatives, *Marine Mammal Amendments: Hearings before the Subcommittee on Fisheries and Wildlife Conservation and the Environment of the Committee on Merchant Marine and Fisheries,* 94th Cong., 2nd sess. (April 30, and May 4, 20, 21, and 24, 1976): 78, 99, 167–68.

49. House of Representatives, *Marine Mammal Amendments,* 67, 68, 71, 74, 76.

50. House of Representatives, *Marine Mammal Amendments,* 62, 64, 161, 164.

51. ZooAct Board of Directors Meeting Minutes (May 14, 1976); Leggett quoted in *ZooAction* (September 1976), ZooAct, Inc., Box 28, AZA Record 96-024.

52. Robert Wagner, "Brief Legislative Report," *AAZPA Newsletter* 16, no. 5 (May 1975): 4.

53. "Legislation," *AAZPA Newsletter* 17, no. 1 (January 1976): 2; Robert Wagner, "AAZPA Executive Director Report," *AAZPA Newsletter* 18, no. 10 (October 1977): 3; Robert Wagner, "AAZPA Executive Director Report," *AAZPA Newsletter* 18, no. 11 (November 1977): 13.

54. "Standards for the Care, Maintenance, and Transportation of Marine Mammals," *AAZPA Newsletter* 19, no. 8 (August 1978): 6.

55. George Steele, "Washington Legislative and Regulatory Zoological Highlights—1979," (September 29, 1979), Steele and Utz 1979, Box 38, AZA Record 96-024; Robert Wagner, "AAZPA to be Involved in USDA Training Regarding Marine Mammal Regulations," *AAZPA Newsletter* 20, no. 7 (July 1979): 5; see also G. Steele, "Expert Marine Mammal Panel Established to Review Marine Mammal Regulations," *AAZPA Newsletter* 22, no. 2 (February 1981): 6.

56. "Law Enforcement by USDA's Animal Care Staff," *AWIIR* 21, no. 4 (1972); "Law Enforcement by USDA's Animal Care Staff," *AWIIR* 24, no. 3 (1975): 3; Statement of Sophie F. Danforth, Chairman of the Board, Rhode Island Zoological Society, Senate Subcommittee on the Smithsonian Institution of the Committee on Rules and Administration, *Federal Assistance for Zoos and Aquariums: Hearing before the Subcommittee on the Smithsonian Institution of the Committee on Rules and Administration,* 93rd Cong., 2nd sess., 1974, 32.

57. Senate, *Federal Assistance for Zoos and Aquariums,* 24, 26, 30, 31 49, 51.

58. Statements of George Speidel, Director, Milwaukee County Zoological Park, and Dion A. Albach, Director, Mesker Park Zoo, Senate, *Federal Assistance for Zoos and Aquariums,* 24–25, 48, 84.

59. Senate, *Federal Assistance for Zoos and Aquariums,* 24, 27, 32, 42, 48, 52, 56. The final two comments were from Philip W. Ogilvie, Executive Director, Portland Zoological Gardens, and Ronald T. Reuther, Director, Philadelphia Zoological Garden.

60. Senate, *Federal Assistance for Zoos and Aquariums,* 25, 59.

61. "Grantmanship Program Proves Successful," *AAZPA Newsletter* 19, no. 6 (June 1978): 9–11.

62. On CETA, see Theodore Lowi, *The End of Liberalism: The Second Republic of the United States* (New York: W. W. Norton, 1979), 89–90.

63. "Note to Zoo Directors," *AAZPA Newsletter* 16, no. 3 (March 1975): 5; "Federal Grant Provides 30 Employees to Overton Park Zoo and Aquarium," *AAZPA Newsletter* 16, no. 7 (July 1975); "Federal Job Funding Applied to Zoo," *AAZPA Newsletter* 16, no. 4 (April 1975): 7.

64. Robert Wagner to AAZPA Board of Directors, May 23, 1978, NASORLO, Box 21, AZA Record 96-024.

65. "Notice to Zoological Park and Aquarium Administrators," *AAZPA Newsletter* 17, no. 10 (October 1976): 1; "Franklin Park Zoo Director to Receive Grant," *AAZPA Newsletter* 18, no. 10 (October 1977): 10; "Community Development Funds Received," *AAZPA Newsletter* 19, no. 3 (March 1978): 14; "BOR Grant Received," *AAZPA Newsletter* 19, no. 6 (June 1978): 14; Paul Pritchard (USDI) to Robert Wagner, December 3, 1979, AAZPA Legislation Committee 1979, Box 2, AZA Record 96-024.

66. Museum Services Act, Box 21, AZA Record 96-024; *AAZPA Newsletter* 23, no. 12 (December 1977): 18; "News From Washington," *AAZPA Newsletter* 23, no. 3 (March 1977): 3; "AAZPA Executive Director Report," *AAZPA Newsletter* 14, no. 10 (October 1978): 3; "Announcements," *AAZPA Newsletter* 20, no. 10 (October 1979): 21.

67. See, for example, "Positions Available," *AAZPA Newsletter* 17, no. 4 (April 1976): 12, and *AAZPA Newsletter* 16, no. 3 (March 1975): 15.

68. See, for example, "Positions Available," *AAZPA Newsletter* 17, no. 4 (April 1976): 12; and "Positions Wanted," *AAZPA Newsletter* 26, no. 2 (February 1975): 16.

69. "Positions Available," *AAZPA Newsletter* 16, no. 3 (March 1975): 15; "Position Directory," *AAZPA Newsletter* 23, no. 3 (March 1982): 16.

70. J. Beck, "NEH Names Consultant," *AAZPA Newsletter* 19, no. 5 (May 1978): 5; B. Serrell, "Grant Awarded," *AAZPA Newsletter* 14, no. 6 (June 1978): 14.

71. "Zoo Educational Material Developed for Schools," *AAZPA Newsletter* 18, no. 11 (November 1977): 17.

72. Untitled Letter, *AAZPA Newsletter* 20, no. 6 (June 1979): 1.

73. ZooAct, Inc., Box 28, AZA Record 96-024.

74. William Conway to George Steele, February 3, 1976, ZooAct, Inc., Box 28, AZA Record 96-024.

75. Conway to Steele (emphasis added); statement of John Grandy (Defenders of Wildlife), U.S. House, "Endangered Species Oversight Hearings before the Subcommittee on Fisheries and Wildlife Conservation and the Environment of the Committee on Merchant Marine and Fisheries," 94th Cong., 1st sess., October 1, 2, and 6, 1975, 66–89.

76. ZooAct Board of Directors Meeting Minutes (May 14, 1976), ZooAct, Inc., Box 28, AZA Record 96-024; Paul Chaffee to George Steele, October 31, 1975, Friends of Animals, Box 17, AZA Record 96-024.

77. Wagner to AAZPA Board, October 12, 1977; Wagner and William Meeker, An Open Letter to All Zoological Park/Aquarium Directors (1977); Meeker to Institution/Society Chief Administrators, October 6, 1977 (all in Members Comments 1978 Legislative Program, Box 38, AZA Record 96-024).

78. George Steele, billing statement, July 9, 1979, Steele and Utz 1979, Box 38, AZA Record 96-024.

79. Paul Chaffee to AAZPA Board of Directors, June 25, 1979, Steele and Utz 1979, Box 38, AZA Record 96-024.

80. George Rabb to Paul Chaffee, June 28, 1979, Steele and Utz 1979, Box 38, AZA Record 96-024.

81. "Statement of George Steele," June 5, 1979, AAZPA Legislative Committee 1971–1980, Box 2; George Steele to Edward Maruska, August 24, 1979, Steele and Utz 1979, Box 38; Robert Wagner to Edward Maruska, September 14, 1979, Wildlife Conservation 1979, Box 38, AZA Record 96-024.

82. Wagner to Bill Conway, September 23, 1977, Members Comments 1978 Legislative Program, Box 38, AZA Record 96-024.

83. Robert Wagner, "AAZPA Executive Director Report," *AAZPA Newsletter* 21, no. 1 (1980): 3; *AAZPA Newsletter* 20, no. 12 (1979): 8, 15.

84. Robert Wagner, "Reflections upon 1976," *AAZPA Newsletter* 18, no. 1 (January 1977): 18; Gordon Hubbell, untitled letter to members, *AAZPA Newsletter* 18 no. 3 (March 1977): 1; *AAZPA Newsletter* 20, no. 12 (December 1979): 8, 15; Robert Wagner, "AAZPA Executive Director Report," *AAZPA Newsletter* 20, no. 3 (March 1977): 3.

85. Ed Maruska, "A Message From the President," *AAZPA Newsletter* 20, no. 3 (March 1979): 1.

4—Species Preservation

1. "Information on AAZPA's Species Survival Plan," *AAZPA Newsletter* 21, no. 12 (December 1980): 4–5; Tom Foose, "Conservation Coordinator's Report," *AAZPA Newsletter* 22, no. 4 (April 1981): 8. Foose held the first AAZPA staff position dedicated solely to conservation: the Conservation Coordinator, created in 1980.

2. Tom Foose, "Man and Biosphere Genetics Symposium and Workshop," *AAZPA Newsletter* 22, no. 11 (November 1981): 14–15; American Committee for International Conservation [folder], Box 12, AZA Record 96-024.

3. For example, the Los Angeles Zoo, working with the Zimbabwe government, SAVE, and the AAZPA, acquired two black rhinos. Tom Foose, "Another Pair of Black Rhino on the Ark," *AAZPA Newsletter* 24, no. 1 (January 1983): 4.

4. Robert Bendiner, *The Fall of the Wild, The Rise of the Zoo* (New York: Elsevier-Dutton, 1982); Gordon Woodruffe, *Wildlife Conservation and the Modern Zoo* (Hindhead, England: Saiga Publishing, 1981).

5. "Euthanasia of Zoo/Aquarium Specimens" (August 1, 1981), Surplus Animal Fact Finding Committee, Box 4, AZA Record 99-102; Tom Foose, "Conservation Coordinator's Report," *AAZPA Newsletter* 22, no. 9 (September 1981): 8–9.

6. "AAZPA Board of Directors Meeting," *AAZPA Newsletter* 22, no. 11 (November 1981): 4; "Long-Range Plan," *AAZPA Newsletter* 27, no. 11 (November 1986): 5.

7. "Condensed Minutes, Board of Directors Meeting," *AAZPA Newsletter* 20, no. 12 (December 1979): 12; D. Meritt, "Mexican Wolf Recovery Team Meeting Held," *AAZPA Newsletter* 20, no. 12 (December 1979): 26. "Annual Membership Business Meeting," *AAZPA Newsletter* 24, no. 11 (November 1983): 12. A Mexican wolf SSP was created in 1993.

8. Roland Smith, "Red Wolf," *AAZPA Newsletter* 27, no. 6 (June 1986): 6; Roland Smith, "Red Wolf," *AAZPA Newsletter* 30, no. 1 (January 1989): 5; "AAZPA Board of Directors Meeting," *AAZPA Newsletter* 27, no. 5 (May 1986): 7; "AAZPA Board of Directors Meeting," *AAZPA Newsletter* 27, no. 11 (November 1986): 13.

9. Fish and Wildlife Service, "Captive-Bred Wildlife Regulation," *Federal Register* 57, no. 4 (January 7, 1992): 548–52 (quotation from 549). Charlie Dane, "Permit Processing for Exports of SSP Animals," *AAZPA Newsletter* 30, no. 1 (January 1989): 11–12.

10. "Recent GAO Report Issued on Endangered Species Act Programs," *AAZPA Newsletter* 30, no. 4 (April 1989): 14.

11. The earliest versions of this legislation reported in the *Newsletter* were HR 3010 (1987) and HR 1704 (1989). This "enthusiastic" quotation comes from "AAZPA Board of Directors Meeting," *AAZPA Newsletter* 30, no. 5 (May 1989): 7.

12. HR 4335 (1988) and HR 1268 (1989) will be considered here.

13. James Scheuer, opening comments, Hearings before the Subcommittee on Natural Resources, Agricultural Research and Environment, *The National Biological Diversity Conservation and Environmental Research Act,* 100th Cong., 2nd sess., June 9 and 30, 1988, 1–3.

14. Statement of Theodore Reed (Director, National Zoological Park), House Subcommittee on Fisheries and Wildlife Conservation, *Fish and Wildlife Legislation Part I: Hearings before the House Subcommittee on Fisheries and Wildlife Conservation,* 90th Cong., 1st sess., October 4, 1967, 41–60.

15. AAZPA letter to Scheuer (July 19, 1988), Hearings before the Subcommittee on Natural Resources, Agricultural Research and Environment, *The National Biological Diversity Conservation and Environmental Research Act,* 100th Cong., 2nd sess., June 9 and 30, 1988, 345–48; AAZPA letter to Scheuer (July 24, 1989), Hearings before the Subcommittee on Natural Resources, Agricultural Research and Environment, *H.R. 1268—The National Biological Diversity Conservation and Environmental Research Act,* 100th Cong., 1st sess., May 17, 1989, 204–5. An AAZPA member had testified about the value of SSPs at an earlier (May 28, 1987) House Committee on Science, Space and Technology hearing on biological diversity.

16. Sec. 8 (8) of HR 1268.

17. Robert Wagner to Stephan [*sic*] Graham, December 14, 1976, Detroit Zoo, Box 16, AZA Record 96-024 [Graham later changed the spelling of his first name to "Steve"]; "An AAZPA Zoo Surplus Guideline," *AAZPA Newsletter* 19, no. 11 (November 1978): 14–18; Surplus Animal Policy 1979 [folder], Box 38, AZA Record 96-024.

18. Teresa Nelson to Glenn Manuel, August 16, 1980, AAZPA Animal Welfare Committee 1980, Box 1, AZA Record 96-024; Theresa Nelson to Steve Graham, January 4, 1980, AAZPA Animal Welfare Committee 1981, AZA Record 96-024.

19. Sue Pressman to Steve H. Graham, November 12, 1981, AAZPA Animal Welfare Committee 1981, Box 1, AZA Record 96-024; Steve H. Graham, "Roadside Zoo Committee Annual Activity Report," 1981, Box 1, AZA Record 96-024.

20. Steve H. Graham, "Roadside Zoo Committee Annual Activity Report," 1981; letters from roadside zoos appear in the Animal Welfare Committee files, Box 1, AZA Record 96-024.

21. Robert Wagner, "Executive Director Report," *AAZPA Newsletter* 22, no. 4 (April 1981): 3.

22. Theodore H. Reed to Robert Wagner, December 2, 1981, Box 1, AZA Record 96-024.

23. Memo to AAZPA Board of Directors and Ethics Board from Robert Wagner, December 23, 1981, Box 1, AZA Record 96-024.

24. Gerald Lentz to AAZPA Board, December 10, 1981; George Felton to Robert Wagner, December 18, 1981, AAZPA Animal Welfare Committee 1981, Box 1, AZA Record 96-024.

25. Tyson Smith to Bob Wagner, October 5, 1984, Box 28, AZA Record 96-024.

26. Steve Graham, "Animal Welfare Committee Annual Activity Report," August 2, 1982; Sue Pressman to Steve Graham, February 3, 1982, AAZPA Animal Welfare Committee 1982, Box 1, AZA Record 96-024; Pressman to Graham, February 9, 1983; Ronald Blakely to Paul Chaffee, January 24, 1983, AAZPA Animal Welfare Committee 1983, Box 1, AZA Record 96-024.

27. Doris Applebaum, "Euthanasia of 3 Siberian Tigers at the Detroit Zoo," *AAZPA Newsletter* 23, no. 12 (December 1982): 4–7.

28. Applebaum, "Euthanasia," 4. Quotation about difficulty of decision comes from "Brief in Support of Motion" filed in *Krescentia M. Doppleberger and Fund For Animals vs. City of Detroit, The Detroit Zoological Society, and Steve Graham,* Wayne County Circuit Court, no. 82-234592-cz.

29. Applebaum, "Euthanasia," 4.

30. Applebaum, "Euthanasia," 5.

31. *Doppleberger vs. City of Detroit;* "The Lady and the Tigers: An Unlikely Champion," *Detroit Free Press,* September 30, 1982 (Newsbank.com, accessed March 22, 2004).

32. "Complaint," *Doppleberger vs. City of Detroit;* Cleveland Amory, *Man Kind? Our Incredible War on Wildlife* (New York: Harper and Row, 1974); "Detroit Zoo's Tigers Get A Court Reprieve," *Detroit Free Press,* September 25, 1982 (Newsbank.com, accessed March 22, 2004); "Order Continuing Preliminary Injunction Issued by the Hon. Joseph B. Sullivan with Stipulations Thereto," September 24, 1982, *Doppleberger vs. City of Detroit.*

33. Applebaum, "Euthanasia," 5; Court of Appeals, October 25, 1982 (emphasis added); "City of Detroit's Supplemental Brief to its Original Motion for Summary Judgment," November 22, 1982; "Closing Arguments," November 3, 1982, *Doppleberger vs. City of Detroit.*

34. Applebaum, "Euthanasia," 6; "Detroit Zoo's Tigers Get a Court Reprieve"; Robert Wagner, "Executive Director's Report," *AAZPA Newsletter* 23, no. 12 (December 1982): 3; Draft of "Euthanasia of Zoo/Aquarium Specimens" (August 1, 1991), Box 4, AZA Record 96-024.

35. "Proceedings," November 3, 1982, 97–101, *Doppleberger vs. City of Detroit;* Dorris Applebaum, "Detroit Zoo Siberian Tiger Update," *AAZPA Newsletter* 24, no. 1 (January 1983): 6.

36. "Court Proceedings," September 24, 1982, *Doppleberger vs. City of Detroit;* "Proceedings," November 3, 1982, 102, *Doppleberger vs. City of Detroit;* "Closing Arguments," November 3, 1982, 3, *Doppleberger vs. City of Detroit;* handwritten note to Wagner, January 30, 1984, Detroit Zoo, Box 16, AZA Record 96-024; Robert Wagner, "Executive Director's Report," *AAZPA Newsletter* 23, no. 12 (December 1982): 3. The Great Lakes Regional AAZPA Conference also devoted a panel to euthanasia in 1983.

37. Sue Pressman to Judge Paul S. Teranes, October 29, 1982, Box 1, AZA Record 96-024; Robert Wagner to Sue Pressman, December 22, 1982, AAZPA Animal Welfare Committee 1982, Box 1, Record 96-024.

38. Pressman to Teranes; Robert Wagner to Sue Pressman, December 22, 1982; Wagner to Steve Taylor, December 22, 1982, Animal Welfare Committee 1982, Box 1, AAZPA Record 96-024.

39. Doris Applebaum, "Euthanasia of Four Healthy Tigers at the Detroit Zoo Reported," *AAZPA Newsletter* 26, no. 6 (June 1985): 14–15; on the API's position on SSPs, see Steve Taylor to Robert Wagner, December 13, 1982, Animal Protection Institute, Box 12, AZA Record 96-024.

40. Belton P. Mouras to Linda Boyd, August 2, 1985, Animal Protection Institute, Box 12, AZA Record 96-024.

41. Ken Kawata to AWC members, September 10, 1984, AAZPA Animal Welfare Committee 1984; Steve Graham to Ken Kawata, January 14, 1986, Surplus Animal Welfare Questionnaire 1986, Box 1, AZA Record 96-024.

42. *Tim Jones v. William G. Gordon,* 792 F.2d 821; 1986 U.S. App. Lexis 26197.

43. Robert Wagner, "Executive Director's Report," *AAZPA Newsletter* 24, no. 8 (August 1983): 3.

44. *Tim Jones v. William G. Gordon,* 792 F.2d 821; 1986 U.S. App. Lexis 26197, 2.

45. Belton P. Mouras to Lanny Cornell, November 1, 1983, Box 12, AZA Record 96-024.

46. Kristin Vehrs, "Public Hearing on Sea World Killer Whale Application," *AAZPA Newsletter* 24, no. 9 (September 1983): 7.

47. George F. Will, "The Orcas of Sea World," *Newsweek,* August 29, 1983, Box 10, AZA Record 96-024.

48. *Tim Jones v. William G. Gordon;* Robert Hunter, *Warriors of the Rainbow: A Chronicle of the Greenpeace Movement* (New York: Holt, Rinehart and Winston, 1979), 128; Vehrs, "Public Hearing," 6–7.

49. Sea World, Inc., Application for Permit to Take Killer Wales, March 7, 1983, AZA, Box 10, Record 96-024. William Braker to Robert Brumsted, August 15, 1983, Box 10, AZA Record 96-024.

50. Vehrs, "Public Hearing," 6–8; "Sea World Faces Fight on Whale Hunt," *Detroit Free Press,* August 17, 1983 (Newsbank.com, accessed March 22, 2004).

51. "Killer Whale Tests Approved," *Washington Post,* November 3, 1983 (Newsbank.com, accessed March 22, 2004); "Sea World Permitted to Test Whales," *Philadelphia Enquirer,* November 2, 1983 (Newsbank.com, accessed March 22, 2004).

52. *Tim Jones v. William G. Gordon.*

53. *Tim Jones v. William G. Gordon.*

54. Kristin Vehrs, "Senate Holds Hearing on the Reauthorization of the Marine Mammal Protection Act," *AAZPA Newsletter* 29, no. 6 (June 1988): 11; Kristin Vehrs, "Marine Mammal Protection Act Changes Become Law," *AAZPA Newsletter* 30, no. 1 (January 1989): 9–11.

55. Kristin Vehrs, "NMFS Holds Hearing on Status of Tursiops Truncatus," *AAZPA Newsletter* 29, no. 6 (June 1989): 9; "Board of Directors Meeting," *AAZPA Newsletter* 30, no. 11 (November 1989): 7; "AAZPA Attends NMFS Working Sessions on Marine Mammal Permit Process," *Communiqué* (February 1990): 11; Kristin Vehrs, "Marine Mammal Public Display Reform Act," *Communiqué* (June 1990): 6.

56. CITES, Article III, 3(c).

57. George Schaller, *The Last Panda* (Chicago: University of Chicago Press, 1993), 15.

58. Intervening Defendants' Memorandum of Points of Authorities in Opposition to Motion for Preliminary Injunction, June 1, 1988, 4.

59. Schaller, *The Last Panda,* 238.

60. Robert Wagner, "Highlights of AAZPA/WWF Lawsuit" (no date), Giant Panda Task Force Toledo Situation, Box 55, AZA Record 96-024.

61. See "Report on the Giant Panda Task Force" Exhibit D, from Affidavit of Paul K. Voskull, June 1988.

62. Exhibit E, Letter from AAZPA to Bill Dennler (Toledo Zoo Executive Director) April 18, 1988, American Association of Zoological Parks and Aquariums in Plaintiffs' Reply to Oppositions Motion for Preliminary Injunction, Filed June 15, 1988; Schaller, *The Last Panda,* 241; Wagner, "Highlights."

63. *World Wildlife Fund v. Donald P. Hodel,* 1988 U.S. Dist. Lexis 19409.

64. *World Wildlife Fund v. Donald P. Hodel.*

65. Schaller, *The Last Panda,* 241; Exhibit E, Affidavit of Kenneth Berlin, May 12, 1988 (original emphasis).

66. Intervening Defendants' Memorandum of Points and Authorities in Opposition to Motion for Preliminary Injunction, June 1, 1988, 1 (original emphasis).

67. Plaintiff WWF's Memorandum of Points and Authorities in Opposition to Toledo Zoo's Motion to Expedite Discovery or to Defer Further Proceedings, June 6, 1988. The Toledo Zoo had argued that the WWF lacked standing.

68. Intervening Defendants' Memorandum of Points and Authorities in Opposition to Motion for Preliminary Injunction June 1, 1988, 5, 13.

69. Motion of the State of Ohio to Intervene, May 24, 1988, 2.

70. *World Wildlife Fund v. Donald P. Hodel.*

71. Palmer Krantz III to William K. Reilly, January 17, 1989; William K. Reilly to Robert Wagner, January 4, 1989, Box 28, AZA Record 96-024; M. Rupert Cutler to Wagner, May 30, 1989, Defenders of Wildlife, Box 15, AZA Record 96-024.

72. Robert Wagner to AAZPA Board of Directors, December 24, 1987; Robert Wagner memo to the AAZPA Membership Committee June 22, 1987; Russell E. Train to Robert Wagner June 11, 1987, Box 28, AZA Record 96-024.

73. David Herbet to Robert Wagner, October 6, 1986, HSUS 1985–89, Box 18, AZA Record 96-024.

74. Kristin Vehrs to Ed Schmitt, February 14, 1989, HSUS 1985–89, Box 18, AZA Record 96-024; Anna Fesmire, "Outline of 'The Role of the HSUS in

Zoo Reform'" [memo to HSUS Board of Directors], March 27, 1980, and John Grandy to Karen Asis, October 23, 1990, HSUS 1980–1990, Box 18, AZA Record 96-024; James Wyerman to Robert Wagner, July 26, 1990, Defenders of Wildlife, Box 15, AZA Record 96-024; Wagner memo, "HSUS Statement of Policy," January 12, 1989, Legislation Committee 1988–89, Box 55, AZA Record 96-024; Wagner to David Herbet, May 10, 1989, Detroit Zoo, Box 16, AZA Record 96-024; Wagner to Grandy, September 6, 1988, HSUS 1985–89, Box 18, AZA Record 96-024; Ed Schmitt and Kris Vehrs to Krantz and Wagner, May 31, 1989, Legislation Committee 1988–89, Box 55, AZA Record 96-024.

75. Paul Chaffee to Wagner, August 1, 1989, HSUS 1985–89, Box 18, AZA Record 96-024; Grandy article appeared in *The Latham Letter* (Fall 1989); PETA Factsheet, May 23, 1988, PETA, Box 23, AZA Record 96-024.

76. Edward Schmitt, "Effects of Conservation Legislation on the Professional Development of Zoos," *International Zoo Yearbook,* vol. 27 (1988): 3–9; William Conway to Paul Chaffee, April 8, 1981, Box 20, AZA Record 96-024.

77. James Deane to Warren Iliff, February 26, 1987, Defenders of Wildlife, Box 15, AZA Record 96-024. Quotation is title of an article from *Audubon Adventures,* National Audubon Society, Box 21, AZA Record 96-024.

5—Animal Welfare Issues Revisited

1. House Subcommittee on Department Operation, Research and Foreign Agriculture of the Committee of Agriculture, *Review of the U.S. Department of Agriculture's Enforcement of the Animal Welfare Act, Specifically of Animals Used in Exhibition: Hearing before the Subcommittee on Department Operation, Research and Foreign Agriculture of the Committee of Agriculture, House of Representatives,* 102nd Cong. 2nd sess., July 8, 1992, Serial no. 102-75, 2. Testimony from 715–16.

2. *In Defense of Animals v. Cleveland Metroparks Zoo,* 785 F. Supp. 100; 1991 U.S. Dist.

3. Steve Taylor, "From the President," *Communiqué* (August 1992): 13.

4. *Humane Society v. Bruce Babbitt,* 310 U.S. App. D.C. 228; 46 F.3d 93: 1995 U.S. App.; Reply for Appellant: The Humane Society of the United States, November 1994, 13; Brief for Appellee Hawthorn Corporation, November 1994, 19; Kris Vehrs to AAZPA Board of Directors, June 24, 1992, Milwaukee County Zoo, Box 21, AZA Record 96-024; *The Humane Society of the United States v. Bruce Babbitt,* 1995, 310 U.S. App. D.C. 228; 46 F.3d 93; "County Zoo Director Won't Let Boycott Sway Him," *Milwaukee Journal,* December 28, 1990 (Newsbank.com, accessed March 27, 2004); "Lota Flap a Surprise to Some," *Milwaukee Sentinel,* December 29, 1990 (Newsbank.com, accessed March 27, 2004).

5. Regulations quoted in *Humane Society v. Bruce Babbitt.*

6. AAZPA to Deputy Administrator of Veterinary Services, APHIS (March 29, 1973), AAZPA Board of Directors Mid-Year Meeting Minutes (March 29–30, 1973), Box 9, AZA Record 99-102.

7. Department of the Interior, "Policy on Giant Panda Import Permits; Request for Comments," *Federal Register* 57 (June 29, 1992): 28825; "Development of Permit Policy for Import of Giant Pandas; Suspension of Consideration of Giant Panda Import Permit Applications, and a Review of Existing Policy on Giant Panda Import Permits." *Federal Register* 59, no. 85 (May 4, 1994): 23077; Department of the Interior, "Proposed

Policy on Giant Panda Permits," *Federal Register* 60, no. 61 (March 30, 1995): 16487–98.

8. Robert Wagner, "Executive Director's Report," *AAZPA Newsletter* 28, no. 10 (October 1987): 3–4; Interior, "Proposed Policy" (March 30, 1995).

9. Interior, "Proposed Policy" (March 30, 1995); Department of the Interior, "Policy on Giant Panda Permits" *Federal Register* 63, no. 166 (August 27, 1998): 45839.

10. Robert Wagner, "Executive Director's Report," *AAZPA Newsletter* 25, no. 8 (August 1984): 3; Robert Wagner, "Executive Director's Report," *AAZPA Newsletter* 26, no. 4 (April 1985): 3.

11. Interior, "Policy on Giant Panda Permits" (August 27, 1998).

12. "Meeting Minutes," *Communiqué* (December 2002): 29; Steve Olson, "CITES Update," *Communiqué* (December 2004): 15–17.

13. John W. Grandy, "Zoos: A Critical Reevaluation," *HSUS News* (Summer 1992): 12–14; "Good Advice for the Aquarium Industry," *Animal Welfare Institute Quarterly* 40, no. 1 (Spring 1991): 5; "Imprisoned Without Trial" flier (no date), Animal Rights, Box 1, AZA Record 99-102; "Zoo Act Task Force Action Plan" (July 1991), Animal Rights, Box 1, AZA Record 99-102. Burdett Loomis and Allan J. Cigler, "Introduction: The Changing Nature of Interest Group Politics," in Burdett Loomis and Allan J. Cigler, eds., *Interest Group Politics,* 5th ed. (Washington, D.C.: Congressional Quarterly Press, 1998), 22.

14. Gary L. Francione, *Animals, Property, and the Law* (Philadelphia: Temple University Press, 1995).

15. "*In Defense of Animals v. City of Los Angeles & Department of Interior* (Los Angeles Gorilla Transfer)," *Communiqué* (August 1992): 3.

16. *Citizens to End Animal Suffering and Exploitation v. The New England Aquarium,* 836 F. Supp 45; 1993 U.S. Dist. Kristin Vehrs, "Aquarium and PAWS Reach Agreement in Kama Lawsuit," *Communiqué* (November 1993): 3.

17. *Lujan v. Defenders of Wildlife* (1992).

18. *CEASE v. The New England Aquarium.*

19. *The Humane Society of the United States v. Bruce Babbitt.*

20. *Animal Lovers Volunteer Association v. Weinberger,* 765 F.2d 937, 938-38, Ninth Cir.

21. Kristin Vehrs, "AAZPA Comments on Animal Health Protection Act," *Communiqué* (August 1990): 6–7.

22. House Subcommittee on Department Operations, Research, and Foreign Agriculture of the Committee on Agriculture, *Farm Animal and Research Facilities Protection Act of 1989: Hearing Before the Subcommittee on Department Operations, Research, and Foreign Agriculture of the Committee on Agriculture House of Representatives,* 101st Cong. (Washington, D.C.: Government Printing Office, 1990): 1, 29. Lorenz Lutherer, Margaret Simon, and John Orem, *Targeted: The Anatomy of an Animal Rights Attack* (Norman: University of Oklahoma Press, 2000).

23. Peter Singer, "Prologue: Ethics and the New Animal Liberation Movement"; Tom Regan, "The Case for Animal Rights," 13–26; Dale Jamieson, "Against Zoos," 109–17, all in Peter Singer, ed., *In Defense of Animals* (New York: Harper & Row, 1985). Steven M. Wise, *Rattling the Cage: Toward Legal Rights for Animals* (Cambridge, Mass.: Perseus Books, 2000).

24. Jane Fristsch, "Animal Activists Vandalize Zookeeper's Home," *Los Angeles Times* (October 15, 1988): 1, 26; quoted in Bettina Manzo, *The Animal Rights Movement*

in the United States 1975–1990: An Annotated Bibliography (Metuchen, N.J.: Scarecrow Press, 1994), 139–40.

25. Interview with Kevin Walsh, June 30, 2005.

26. Interview with David Scofield, June 31, 2005.

27. *Commonwealth v. Richard Strahan* (1995), No. 94-P-1657, Appeals Court of Massachusetts.

28. Linda Graham Caleca, "Zoo Officals Told to Face Up to Issue of Animal Rights," Box 13, AZA Record 96-024.

29. William Braker to Marine Mammal Coalition members, Legislation, March 15, 1990, Box 56, AZA Record 96-024. Robert Wagner, "Executive Director's Report," *Communiqué* (October 1991): 2; Christen Wemmer and Steven Thompson, "A Short History of Scientific Research in Zoological Gardens," in Christen W. Wemmer, ed., *The Ark Evolving: Zoos and Aquariums in Transition* (Front Royal, Va.: Smithsonian Institution Press, 1996), 73; "H.R. 2407—Farm Animal Research Facilities Protection Act," *Communiqué* (August 1992): 2.

30. House Subcommittee on Department Operation, Research and Foreign Agriculture of the Committee of Agriculture, *Review of the U.S. Department of Agriculture's Enforcement of the Animal Welfare Act, Specifically of Animals Used in Exhibition: Hearing before the Subcommittee on Department Operation, Research and Foreign Agriculture of the Committee of Agriculture, House of Representatives,* 102nd Cong. 2nd sess., July 8, 1992, Serial no. 102-75. (Animal and Plant Health Inspection Service Implementation of the Animal Welfare Act Washington, D.C. Audit Report. No. 33002-0001-Ch, March 16, 1992, p. 1 of report. Page 633 of the Congressional Hearing.) The report prompted Representative Peter Kostmayer (D-PA) to introduce the "Exhibition Animal Protection Act" (HR 3252) to remedy APHIS's lack of enforcement of the Animal Welfare Act.

31. House, *Review of the Animal Welfare Act,* 89, 216–17, 718–19.

32. House, *Review of the Animal Welfare Act,* 333.

33. House, *Review of the Animal Welfare Act,* 9–11, 16–17.

34. The ALDF had challenged the USDA's primate regulations in 1993, but failed to gain standing. See *Animal Legal Defense Fund v. Secretary of Agriculture,* 813 F. Supp. 882 (D.D.C. 1993).

35. *Animal Legal Defense Fund v. Daniel Glickman,* 943 F. Supp. 44; 1996, U.S. Dist.

36. *Animal Legal Defense Fund v. Daniel Glickman,* 1998, 332 U.S. App. D.C. 104; 154 F.3d 426.

37. Brief for Intervenor-Appelle, National Association for Biomedical Research, *ALDF v. Glickman* (June 11, 1997).

38. Specifically, the regulation stated that "Dealers, exhibitors, and research facilities must develop, document, and follow an appropriate plan for environmental enhancement adequate to promote the psychological well-being of nonhuman primates." Quoted in *Animal Legal Defense Fund v. Daniel Glickman* (1996).

39. *Animal Legal Defense Fund v. Daniel Glickman* (1996).

40. *Animal Legal Defense Fund v. Daniel Glickman* (1996).

41. Kristin Vehrs, "AAZPA Files Comments on Proposed Animal Welfare Standards Regarding Nonhuman Primates," *AAZPA Newsletter* 30, no. 9 (September 1989): 12.

42. Kristin Vehrs, "APHIS Publishes Final Rule on the Psychological Well-being for Nonhuman Primates," *Communiqué* (May 1991): 6–7.

43. Friends of the Animal Leaflet, Animal Rights Folder, Box 1, AZA Record 99-102; Donald G. Lindburg," "Zoos and the 'Surplus' Problem," Surplus Animal Task Force, Box 9, AZA Record 99-102; "An AAZPA Zoo Surplus Guideline Preamble," *AAZPA Newsletter* 19, no. 11 (November 1978): 14–15.

44. Randy Fulk, "Surplus Animal Analysis" (November 14, 1990), Surplus Animal Fact Finding Task Force File, Box 4, AZA Record 99-102; "Euthanasia of Zoo/Aquarium Species," Surplus Animal Fact Finding Task Force File, Box 4, AZA Record 99-102.

45. "Euthanasia."

46. "Euthanasia."

47. "Euthanasia."

48. "Euthanasia."

49. "AAZPA Policy Concerning the Disposition of Zoo or Aquarium Animals to Organizations or Individuals Which Allow Hunting or the Breeding of Animals for Hunting," *Communiqué* (January 1993): 5–6.

50. "AAZPA Policy Concerning the Disposition of Zoo or Aquarium Animals."

51. In re Sugarloaf Dolphin Sanctuary, Case No. SE960112, Hr'g Tr. At 1-11140 (Dep't Commerce, NOAA, February 8–11, 1999): 289.

52. In re Sugarloaf Dolphin Sanctuary, Case No. SE960112, Hr'g Tr. At 1-11140 (Dep't Commerce, NOAA, February 8–11, 1999): 805–20.

53. Sugarloaf Dolphin Sanctuary, Hr'g Tr.: 308, 354.

54. Sugarloaf Dolphin Sanctuary, Hr'g. Tr.: 313.

55. Sugarloaf Dolphin Sanctuary, Hr'g Tr.: 375–76.

56. In re Sugarloaf Dolphin Sanctuary, Case No. SE960112FM/V (Dep't Commerce, NOAA June 8, 1999).

57. Sugarloaf Dolphin Sanctuary, Hr'g Tr.: 1103.

58. Sugarloaf Dolphin Sanctuary Hr'g Tr.: 116, 979; In re Sugarloaf Dolphin Sanctuary, Case No. SE960112FM/V (Dep't Commerce, NOAA June 8, 1999) 11 12; Sugarloaf Dolphin Sanctuary, Hr'g Tr.: 513, 1069, 1021.

59. Sugarloaf Dolphin Sanctuary, Hr'g Tr.: 1062.

60. Sugarloaf Dolphin Sanctuary, Hr'g Tr.: 296–97; David Zucconi to John Grandy, November 28, 1990, HSUS, 1980–1990, Box 18, AZA Record 96-024.

61. William Conway to Paul Chaffee, April 8, 1981, Lobbying, Box 20, AZA Record 96-024.

Conclusion—Elephants and the Trajectory of Zoo Politics

1. "Elephant Owner Admits Guilt in Care Violations," *Milwaukee Journal Sentinel,* March 16, 2004; Elephant Sanctuary in Tennessee, press release, February 14, 2005 (www.elephants.com/pr/lota_214_05.htm, accessed November 5, 2005).

2. "Animal Rights Groups Target Zoo Elephants," *Oregonian,* May 17, 2005 (www.newsreview.info, accessed May 17, 2005); "What's Happening in Your Zoo? Recent Deaths Fuel Movement to End Elephant Exhibits," U.S. Newswire PETA Press Release, May 12, 2005 (news.yahoo.com, accessed May 17, 2005).

3. "Candidates Take Opposing Positions on Zoo Confinement," NBC 4, May 13, 2005 (www.nbc4.tv, accessed May 17, 2005); Dana Bartholomew, "Mayor's Zoo Study Could Have Elephants Packing Their Trunks," *Daily News of Los Angeles,* August 20,

2005 (Newsbank.com, accessed November 4, 2005); "Prosecuters Launch Investigation of Lincoln Park Zoo," Associated Press, May 16, 2005 (www.belleville.com, accessed May 17, 2005); Andrew Hermann, "Pack Up Elephant Exhibit, Alderman Asks Zoo," *Chicago Sun-Times*, March 10, 2005 (www.elephants.com/news, accessed November 5, 2005), "Revised Elephant Ordinance Would Benefit Circus—Sort of," *Chicago Sun-Times*, July 27, 2005 (Newsbank.com, accessed November 4, 2005), and "Zoo Elephants Have No Educational Value, Expert Says," *Chicago Sun-Times*, August 26, 2005 (findarticles.com, accessed November 3, 2005).

4. Daniel Borunda, "Zoo, Sanctuary Advocates Disagree on Fate of Elephants," *El Paso Times*, June 7, 2005 (elpasotimes.com, accessed June 7, 2005); "Elephants Will Remain at the El Paso Zoo," *El Paso Times*, July 27, 2005 (www.elephants.com, accessed November 3, 2005).

5. Christy Strawser, "Zoo, Association Disagree on Elephants in Captivity," *Daily Oakland Press* (www.theoaklandpress.com, accessed July 13, 2004); Hugh McDiarmid, Jr., "Elephant Caravan Still on the Move," *Detroit Free Press*, April 7, 2005 (www.elephants.com, accessed November 5, 2005).

6. "San Francisco to Possibly Ban Elephants from Zoo," November 18, 2005 (www.zoo-talk.com, accessed January 1, 2005); "San Francisco Moves Elephants to Sanctuary," November 28, 2004 (www.zoo-talk.com, accessed January 1, 2005); Mike Taylor, "'Tinkerbelle' Retires to the Foothills," *Calaveras Enterprise*, December 3, 2004 (www.calaverasenterprise.com, accessed December 4, 2004).

7. "Zoo, Association Disagree on Elephants in Captivity"; In Defense of Animals, "Elephant from San Francisco Zoo Euthanized Due to Severe Captivity-Induced Foot Problems," press release, March 25, 2005 (www.elephants.com, accessed November 5, 2005).

8. "AZA Statement Regarding San Francisco Zoo Elephant Decision," June 23, 2004 (www.aza.org, accessed January 1, 2005); "American Zoo and Aquarium Association Announces Elephant Resolution," December 3, 2004 (www.aza.org, accessed January 1, 2005); Hugh McDiarmid, Jr., "Detroit Zoo Could Change Course for Aging Elephants," *Detroit Free Press*, December 16, 2004 (www.freep.com, accessed December 29, 2004); Strawser, "Zoo, Association Disagree."

9. Bill Foster, no title, *Communiqué* (March 2005): 5, 51.

10. "U.S. Adults Agree that Seeing Elephants in Real Life Helps Promote Education and Animal Conservation" and "Elephant Conservation" (www.aza.org, accessed June 2, 2005); Michael Hutchins and William Conway, "What's in a Name: Zoos vs. Sanctuary," *Communiqué* (August 2004): 54–56; Strawser, "Zoo, Association Disagree."

11. Julie Stoiber, "Elephant Turf at Zoo Sparking a Fight," *Philadelphia Enquirer*, September 22, 2005 (Newsbank, accessed November 4, 2005); "Philadelphia Zoo Puts Off Plans for New Elephant Exhibit," October 29, 2005 (www.phillyburbs.com, accessed November 3, 2005).

12. David B. Truman, *The Governmental Process: Political Interest Groups and Public Opinion* (New York: Alfred A. Knopf, 1951). For a discussion of how active minorities often govern interest groups, see 139, 353–59.

13. Robert A. Kagan, *Adversarial Legalism: The American Way of Law* (Cambridge, Mass.: Harvard University Press, 2001).

14. E. E. Schattschneider, *The Semisovereign People: A Realist's View of Democracy in America* (New York: Holt, Rinehart, and Winston, 1960).

15. Robert Wagner, e-mail to the authors, April 6, 2004.

16. James Q. Wilson, *The Politics of Regulation* (New York: Basic Books, 1980); James Q. Wilson, *Bureaucracy: What Government Agencies Do and Why They Do It* (New York: Basic Books, 1989), 179.

17. See Gary L. Francione, *Animals, Property and the Law* (Philadelphia: Temple University Press, 1995), 230–31.

18. Francione, *Animals, Property and the Law,* 232. Howard Latin, "Regulatory Failure, Administrative Incentives, and the New Clean Air Act," 21 *Environmental Law,* 1647 (1991).

19. Carol Morello, "Baghdad's Needs Extend to Zoo Residents," *Washington Post,* April 30, 2003 (NewsBank.com, accessed June 26, 2004).

Selected Bibliography

Archival and Other Primary Sources

AAZPA Newsletter (1975–1991)

Animal Welfare Institute. *Information Report and AWI Quarterly Compilation,* Vol. 1, 1951–1980.

Animal Welfare Institute. *Information Report and AWI Quarterly Compilation,* Vol. 2, 1980–2001.

AZA Archives, Smithsonian Institution

Communiqué (1992–present)

Defenders of Wildlife News (1964 and 1968)

Court Cases

Animal Legal Defense Fund v. Daniel Glickman, 943 F. Supp. 44 (U.S. Dist. 1996).

Animal Protection Institute of America v. Robert Mosbacher, 799 F. Supp. 173 (U.S. Dist. 1992).

Citizens to End Animal Suffering and Exploitation, Inc. v. The New England Aquarium, 836 F. Supp 45 (U.S. Dist. 1993).

Blanche Guzzi v. New York Zoological Society, 192 A.D. 263; 182 N.Y.S. 257 (N.Y. App. Div. 1920).

In Defense of Animals v. Cleveland Metroparks Zoo, 785 F. Supp. 100 (U.S. Dist. 1991).

In re O'Barry, Docket No. SE960112FM/V (Dep't Commerce, NOAA, June 8, 1999).

Krescentia M. Doppleberger and Fund For Animals vs. City of Detroit, The Detroit Zoological Society, and Steve Graham, (Wayne County Circuit Court, no. 82-234592-cz. 1982).

The Humane Society of the United States v. Bruce Babbitt, 310, 228; 46 F.3d 93, (U.S. App. D.C. 1995).

Tim Jones v. William G. Gordon, 792 F.2d 821 (U.S. App. 1986).

United States Department of Commerce, National Oceanic and Atmospheric Administration, Office of the Administrative Law Judge, Norfolk, Virginia. *In the Matter of Richard O'Barry, Lloyd A. Good, III, Sugarloaf Dolphin Sanctuary, Inc., The Dolphin Project, Inc.* SE960112FM/V (June 1999).

World Wildlife Fund v. Donald P. Hodel, 1988 (U.S. Dist. 1988).

Government Documents

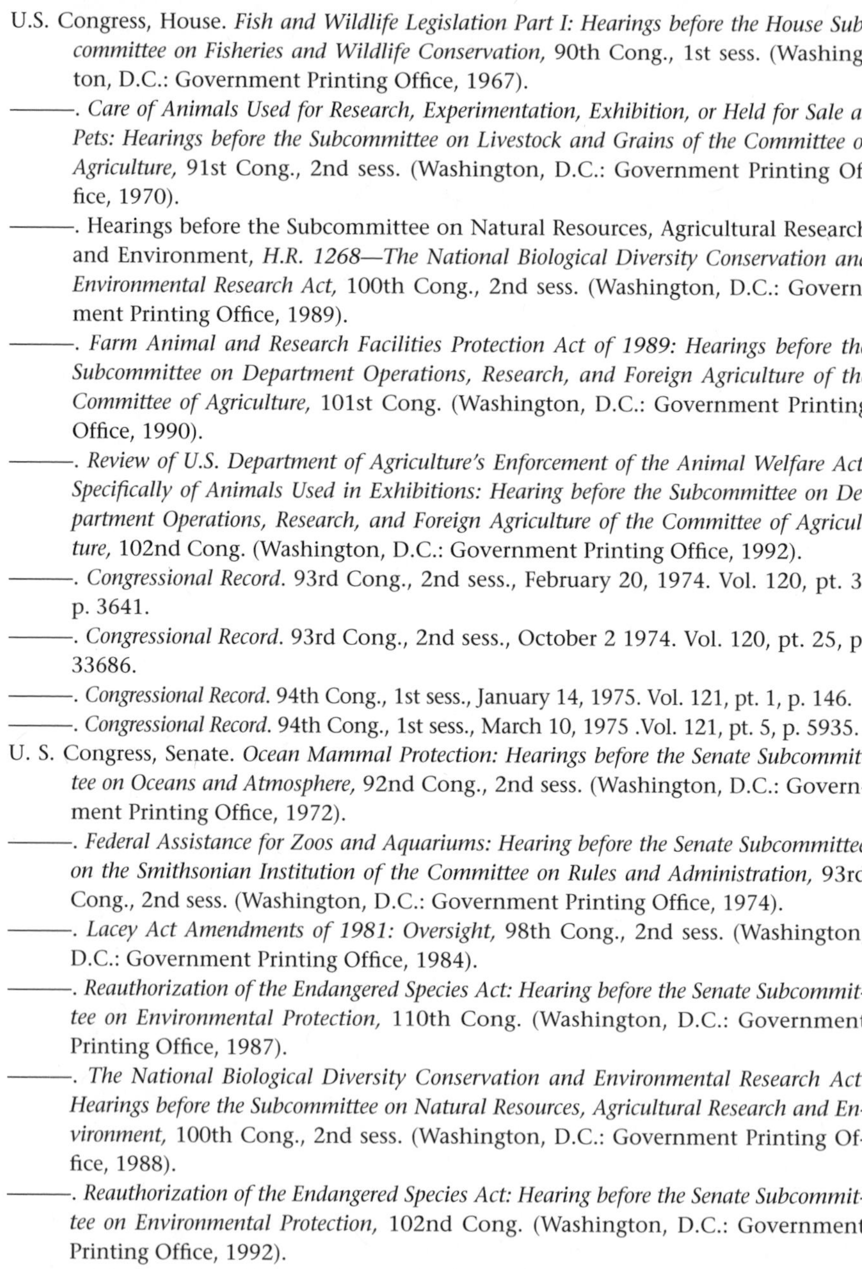

U.S. Congress, House. *Fish and Wildlife Legislation Part I: Hearings before the House Subcommittee on Fisheries and Wildlife Conservation,* 90th Cong., 1st sess. (Washington, D.C.: Government Printing Office, 1967).

———. *Care of Animals Used for Research, Experimentation, Exhibition, or Held for Sale as Pets: Hearings before the Subcommittee on Livestock and Grains of the Committee of Agriculture,* 91st Cong., 2nd sess. (Washington, D.C.: Government Printing Office, 1970).

———. Hearings before the Subcommittee on Natural Resources, Agricultural Research and Environment, *H.R. 1268—The National Biological Diversity Conservation and Environmental Research Act,* 100th Cong., 2nd sess. (Washington, D.C.: Government Printing Office, 1989).

———. *Farm Animal and Research Facilities Protection Act of 1989: Hearings before the Subcommittee on Department Operations, Research, and Foreign Agriculture of the Committee of Agriculture,* 101st Cong. (Washington, D.C.: Government Printing Office, 1990).

———. *Review of U.S. Department of Agriculture's Enforcement of the Animal Welfare Act, Specifically of Animals Used in Exhibitions: Hearing before the Subcommittee on Department Operations, Research, and Foreign Agriculture of the Committee of Agriculture,* 102nd Cong. (Washington, D.C.: Government Printing Office, 1992).

———. *Congressional Record.* 93rd Cong., 2nd sess., February 20, 1974. Vol. 120, pt. 3, p. 3641.

———. *Congressional Record.* 93rd Cong., 2nd sess., October 2 1974. Vol. 120, pt. 25, p. 33686.

———. *Congressional Record.* 94th Cong., 1st sess., January 14, 1975. Vol. 121, pt. 1, p. 146.

———. *Congressional Record.* 94th Cong., 1st sess., March 10, 1975 .Vol. 121, pt. 5, p. 5935.

U. S. Congress, Senate. *Ocean Mammal Protection: Hearings before the Senate Subcommittee on Oceans and Atmosphere,* 92nd Cong., 2nd sess. (Washington, D.C.: Government Printing Office, 1972).

———. *Federal Assistance for Zoos and Aquariums: Hearing before the Senate Subcommittee on the Smithsonian Institution of the Committee on Rules and Administration,* 93rd Cong., 2nd sess. (Washington, D.C.: Government Printing Office, 1974).

———. *Lacey Act Amendments of 1981: Oversight,* 98th Cong., 2nd sess. (Washington, D.C.: Government Printing Office, 1984).

———. *Reauthorization of the Endangered Species Act: Hearing before the Senate Subcommittee on Environmental Protection,* 110th Cong. (Washington, D.C.: Government Printing Office, 1987).

———. *The National Biological Diversity Conservation and Environmental Research Act: Hearings before the Subcommittee on Natural Resources, Agricultural Research and Environment,* 100th Cong., 2nd sess. (Washington, D.C.: Government Printing Office, 1988).

———. *Reauthorization of the Endangered Species Act: Hearing before the Senate Subcommittee on Environmental Protection,* 102nd Cong. (Washington, D.C.: Government Printing Office, 1992).

———. *Rhino and Tiger Conservation Hearing before the Subcommittee of Fisheries Conservation, Wildlife and Oceans,* 105th Cong. (Washington, D.C.: Government Printing Office, 1998).

———. *Congressional Record.* 93rd Cong., 1st sess., June 20, 1973. Vol. 119, pt. 16, p. 20441.

———. *Congressional Record.* 93rd Cong., 2nd sess., September 25, 1974. Vol. 120, pt. 24, p. 32560.

———. *Congressional Record.* 93rd Cong., 2nd sess., October 3, 1974. Vol. 120, pt. 25, p. 33799.

———. U.S. Department of Interior. "Policy on Giant Panda Import Permits; Request for Comments," *Federal Register* 57 (June 29, 1992): 28825.

———. "Development of Permit Policy for Import of Giant Pandas; Suspension of Consideration of Giant Panda Import Permit Applications, and a Review of Existing Policy on Giant Panda Import Permits." *Federal Register* 59, no. 85 (May 4, 1994): 23077.

———. "Proposed Policy on Giant Panda Permits" *Federal Register* 60, no. 61 (March 30, 1995): 16487–98.

———. "Policy on Giant Panda Permits," *Federal Register* 63, no. 166 (August 27, 1998): 45839.

———. San Diego Zoological Society's application for a CITES import permit issued September 9, 2002. Fish and Wildlife Service.

Books and Articles

Batten, Peter. *Living Trophies* (New York: Thomas Y. Crowell, 1976).

Bekoff, Marc, and Jan Nystrom. "The Other Side of Silence: Rachel Carson's View of Animals," *Zygon* 39, no. 4 (December 2004): 861–83.

Bendiner, Robert. *The Fall of the Wild, The Rise of the Zoo* (New York: Elsevier-Dutton, 1982).

Bosso, Christopher. *Pesticides and Politics: The Life Cycle of a Public Issue* (Pittsburgh: University of Pittsburgh Press, 1987).

Bracken, David. "Recent Escapes From U.S. Zoos Raise Questions about Safety," *St. Louis Post-Dispatch,* August 24, 2003.

Bridges, William. *Gathering of Animals: An Unconventional History of the New York Zoological Society* (New York: Harper and Row, 1974).

Brisk, Fay. "Animals and Airlines." In Emily Leavitt ed., *Animals and Their Legal Rights: A Survey of American Laws from 1641 to 1990* (New York: Animal Welfare Institute, 1990), 106–12.

Brown, Michael, and John May. *The Greenpeace Story* (New York: Dorling Kindersley, 1991).

Burgess, Bonnie. *Fate of the Wild: The Endangered Species Act and the Future of Biodiversity* (Athens: University of Georgia Press, 2001).

"Can Zoos Be Humane?" *Parade Magazine* (February 19, 1984): 12–13.

Cohn, Jeffrey. "Decisions at the Zoo," *BioScience* 42, no. 9 (October 1992): 654–59.

Costain, W. Douglas, and James P. Lester. "The Evolution of Environmentalism." In James P. Lester, ed., *Environmental Politics and Policy: Theories and Evidence* (Durham, N.C.: Duke University Press, 1995), 15–38.

"County Zoo Director Won't Let Boycott Sway Him," *Milwaukee Journal,* December 28, 1990.

Czech, Brian, and Paul R. Krausman. *The Endangered Species Act: History, Conservation, Biology, and Public Policy* (Baltimore: Johns Hopkins University Press, 2001).

"Detroit Zoo's Tigers Get a Court Reprieve," *Detroit Free Press,* September 25, 1982.

"Elephant Owner Admits Guilt in Care Violations," *Milwaukee Journal Sentinel,* March 16, 2004.

Etter, Carolyn, and Don Etter. *The Denver Zoo: A Centenial History* (Boulder, Colo.: Rinehart, 1995).

Francione, Gary L. *Animals, Property and the Law* (Philadelphia: Temple University Press, 1995).

Fristsch, Jane. "Keepers Struck Elephant More than 100 Times Trainer Said," *Los Angeles Times,* May 26, 1988, I:3.

———. "Elephant Beating Was Abuse, Society Reports," *Los Angeles Times,* June 2, 1988, I:26.

———. "Animal Activists Vandalize Zookeeper's Home," *Los Angeles Times*, October 15, 1988, 1:26.

Hanson, Elizabeth. *Animal Attractions: Nature on Display in American Zoos* (Princeton, N.J.: Princeton University Press, 2002).

Hatfield, Mark O. *Against the Grain: Reflections of a Rebel Republican* (Ashland, Ore.: Whitecloud Press, 2000).

Hoage, R. J., and William A. Deiss, eds. *New Worlds, New Animals: From Menagerie to Zoological Park in the Nineteenth Century* (Baltimore: Johns Hopkins University Press, 1996).

Hunter, Robert. *Warriors of the Rainbow: A Chronicle of the Greenpeace Movement* (New York: Holt, Rinehart and Winston, 1979).

Hyson, Jeffrey. "Urban Jungles." Ph.D. dissertation, Cornell University, 1999.

———. "Jungles of Eden: The Design of American Zoos." In Michel Conan, ed., *Environmentalism in Landscape Architecture,* vol. 22, pp. 23–42 (Washington, D.C.: Dumbarton Oaks Research Library Collection, 2000), accessed as a PDF file at www.doaks.org/extexts.html.25, October 10, 2003.

Jamieson, Dale. "Against Zoos." In Peter Singer, ed., *In Defense of Animals* (New York: Harper and Row, 1985), 108–17.

Jasper, James M. "Recruiting Intimates, Recruiting Strangers: Building the Contemporary Animal Rights Movement." In Jo Freeman and Victoria Johnson eds., *Waves of Protest: Social Movements Since the Sixties* (Lanham, Md.: Rowman & Littlefield, 1999), 65–82.

Jasper, James M., and Dorothy Nelkin. *The Animal Rights Crusade: The Growth of a Moral Protest* (New York: Free Press, 1992).

Jasper, Margaret C. *Animal Rights Law* (Dobbs Ferry, N.Y.: Oceana Publications, 1997).

Kagan, Robert A. *Adversarial Legalism: The American Way of Law* (Cambridge, Mass.: Harvard University Press, 2001).

Kaufman, Les, and Kenneth Mallory, eds. *The Last Extinction* (Cambridge, Mass.: MIT Press, 1986).

"Killer Whale Tests Approved," *Washington Post,* November 3, 1983.

Kisling, Vernon, ed. *Zoo and Aquarium History: Ancient Animal Collections to Zoological Gardens* (Boca Raton, Fla.: CRC Press, 2001).

Koebner, Linda. *Zoo: The Evolution of Wildlife Conservation Centers* (New York: Forge, 1994).

"The Lady and the Tigers: An Unlikely Champion," *Detroit Free Press,* September 30, 1982.

Latham, Earl. "The Group Basis of Politics: Notes for a Theory." *American Political Science Review* 46, no. 2 (June 1952): 376–97.

Latin, Howard. "Regulatory Failure, Administrative Incentives, and the New Clean Air Act," *Environmental Law* 21 (1991): 1647–1720.

Leavitt, Emily, ed. *Animals and Their Legal Rights: A Survey of American Laws from 1641 to 1990* (Animal Welfare Institute: New York, 1990).

Lester, James P., ed. *Environmental Politics and Policy: Theories and Evidence* (Durham, N.C.: Duke University Press, 1995).

"Lota Flap a Surprise to Some," *Milwaukee Sentinel,* December 29, 1990.

Lowi, Theodore. *The End of Liberalism: The Second Republic of the United States* (New York: W. W. Norton, 1979).

Lutherer, Lorenz, Margaret Simon, and John Orem, *Targeted: The Anatomy of an Animal Rights Attack* (Norman: University of Oklahoma Press, 2000).

Manzo, Bettina. *The Animal Rights Movement in the United States 1975–1990: An Annotated Bibliography* (Metuchen, N.J.: Scarecrow Press. 1994).

Maple, Terry L., and Erika F. Archibald. *Zoo Man: Inside the Zoo Revolution* (Atlanta: Longstreet, 1993).

Morello, Carol. "Baghdad's Needs Extend to Zoo Residents," *Washington Post,* April 30, 2003.

Niven, Charles D. *History of the Humane Movement* (New York: Transatlantic Arts, 1967).

Norton, Bryan, Michael Hutchins, Elizabeth Stevens, and Terry Maple, eds. *Ethics on the Ark* (Washington D.C.: Smithsonian Institution Press, 1995).

Osborn, Fairfield. "Another Noah's Ark—For New York," *New York Times,* October 27, 1963.

Pacheco, Alex, and Anna Francione. "The Silver Springs Monkeys." In Peter Singer, ed., *In Defense of Animals* (New York: Harper and Row, 1985), 135–47.

Regan, Tom. *The Case for Animal Rights* (Berkeley: University of California Press, 1984).

———. "The Case for Animal Rights." In Peter Singer, ed., *In Defense of Animals* (New York: Harper & Row, 1985), 13–26.

Regenstein, Lewis. *The Politics of Extinction: The Shocking Story of the World's Endangered Wildlife* (New York: MacMillan, 1975).

Schaller, George. *The Last Panda* (Chicago: University of Chicago Press, 1993).

Schattschneider, E. E. *The Semisovereign People: A Realist's View of Democracy in America* (New York: Holt, Rinehart, and Winston, 1960).

"Sea World Faces Fight on Whale Hunt," *Detroit Free Press,* August 17, 1983.

"Sea World Permitted to Test Whales," *Philadelphia Enquirer,* November 2, 1983.

Silverstein, Helena. *Unleashing Righs: Law, Meaning, and the Animal Rights Movement* (Ann Arbor: University of Michigan Press, 1996).

Singer, Peter. *Animal Liberation* (New York: Random House, 1975).

———, ed. *In Defense of Animals* (New York: Harper & Row, 1985).

Truman, David B. *The Governmental Process: Political Interests and Public Opinion* (New York: Knopf, 1951).

Unti, Bernard. *Protecting All Animals: A Fifty Year History of the Humane Society of the United States* (Washington, D.C.: Humane Society Press, 2004).

Walker, Jack L. "The Origin and Maintenance of Interest Groups in America," *American Political Science Review* 77, no. 2 (June 1983): 390–406.

Wemmer, Christen, W. ed. *The Ark Evolving: Zoos and Aquariums in Transition* (Front Royal, Va.: Smithsonian Institution Press, 1995).

Wemmer, Christen, and Steven Thompson. "A Short History of Scientific Research in Zoological Gardens." In Christen W. Wemmer, ed., *The Ark Evolving: Zoos and Aquariums in Transition* (Front Royal, Va.: Smithsonian Institution, 1995).

Willis, Susan. "Looking at the Zoo," *South Atlantic Quarterly* 98, no. 4 (Fall 1999): 669–87.

Wilson, James Q. *Political Organizations* (New York: Basic Books, 1973).

———. *The Politics of Regulation* (New York: Basic Books, 1980).

———. *Bureaucracy: What Government Agencies Do and Why They Do It* (New York: Basic Books, 1989).

Wise, Steven H. *Rattling the Cage: Toward Legal Rights for Animals* (Cambridge, Mass.: Perseus Books, 2000).

Woodruffe, Gordon. *Wildlife Conservation and the Modern Zoo* (Hindhead, England: Saiga Publishing, 1981).

Zeehandelaar, Frederik J., and Paul Sarnoff. *Zeebongo: The Wacky Wild Animal Business* (Englewood Cliffs, N.J.: Prentice-Hall, 1971).

Index

Italicized page numbers indicate photos.